Applications of Neurochemistry

APPLICATIONS OF NEUROCHEMISTRY

Edited by:

Dr. Kunal Bhatt

ARCLER PRESS

www.arclerpress.com

Applications of Neurochemistry

Dr. Kunal Bhatt

Arcler Press
224 Shoreacres Road
Burlington, ON L7L 2H2
Canada
www.arclerpress.com
Email: orders@arclereducation.com

ISBN: 978-1-77469-846-4 (Hardcover)

ABOUT THE EDITOR

Dr Krunal Bhatt (1982) is presently working at Zydus hospital, Vadodara as a chief clinical intensivist and Anaesthesiologist for one year. He did his MBBS from Baroda Medical College. He has worked as a Medical Officer at various corporate hospitals in Vadodara (2006-2009). He joined Indian armed forces and posted to various locations during his tenure (2009-2018). He has served at Pune, Lucknow, Jodhpur, Jammu and Kashmir. He also has obtained Counter Insurgency ops medal and High Altitude serving medals for his serving tenure at line of control in Kashmir. He did his post-graduation in Anaesthesiology from AFMC, Pune, India in 2018. He is Life Member of Indian Medical Association, Indian Society of Anesthesiology, Indian Society of Critical Care Medicine, and European Society of Anaesthesia.

TABLE OF CONTENTS

LIST OF FIGURES

LIST OF TABLES

LIST OF ABBREVIATIONS

5-HT	5-hydroxytryptamine
ACBP	Acyl-CoA binding protein
ACC	Anterior cingulate cortex
ACh	Acetylcholine
AChE	Acetylcholine esterase
ACTH	Adrenal-corticotropic hormone
AD	Alzheimer's disease
ADHD	Attention deficit hyperactivity disorder
AEA	Anandamide
ALIC	Anterior limb of the internal capsule
AMPA	α-amino-3-hydroxy-5-methyl-4-isoxazole propionic acid
ANS	Autonomic nervous system
ANT	Anterior nucleus of the thalamus
AST	Allatostatin
ATP	Adenosine triphosphate
AVP	Arginine vasopressin
BZD	Benzodiazepines
cAMP	Cyclic adenosine monophosphate
CART	Cocaine and amphetamine-regulated transcript
CAT	Choline acetyltransferase
CB1R	Cannabinoid receptors
CCK	Cholecystokinin
cGMP	Cyclic guanosine monophosphate
CN	Cholinergic neuron
CNS	Central nervous system
COMT	Catechol o-methyl transferase
CORT	Cortisol

CRH	Corticotropin-releasing hormone
CSF	Cerebrospinal fluid
DA	Dopamine
DBI	Diazepam-binding inhibitor
DBS	Deep brain stimulation
DC	Deutocerebrum
DCVs	Dense-core vesicles
DSE	Depolarization-induced suppression of excitation
DSI	Depolarization-induced suppression of inhibition
eAC	Embryonic anterior commissures
EB	Ellipsoid body
ENS	Enteric nervous system
ePC	Embryonic posterior commissures
Epi	Epinephrine
ET	Essential tremor
FAAH	Fatty acid amide hydrolase
FB	Fan-shaped body
FSH	Follicle-stimulating hormone
GABA	Gamma-aminobutyric acid
GAD	Glutamic acid decarboxylase
GnRH	Gonadotropin-releasing hormone
GPCRs	G-protein coupled receptors
GPi	Globus pallidus internal
HPLC	High-performance liquid chromatography
IT	Isthmus tract
JH	Juvenile hormone
KP	Kisspeptin
LAL	Lateral accessory lobes
LH	Luteinizing hormone
LTK	Locustatachykinin
MAGL	Monoacylglycerol lipase
MAL	Median accessory lobe
MAO	Monoamine oxidase
MAOIs	Monoamine oxidase inhibitors
MB	MUSHROOM bodies

mGluR	Metabotropic glutamate receptor
MMSE	Mini-mental state examination
MRI	Magnetic resonance imagery
mRNA	Messenger ribonucleic acid
MSH	Melanocyte-stimulating hormone
NE	Norepinephrine
NMDA	N-methyl-D-aspartate
NO	Nitric oxide
NOS	Nitric oxide synthase
NPF	Neuropeptide F
NPVF	Neuropeptide VF
NPY	Neuropeptide Y
NSF	N-ethylmaleimide sensitive factor
NT	Neurotransmitter
OCD	Obsessive-compulsive disorder
OFC	Orbitofrontal cortex
OXY	Oxytocin
PAG	Periaqueductal gray
pc	Posterior commissural
PC	Protocerebrum
PD	Parkinson's disease
PET	Positron emission tomography
PG	Periglomerular
PI	Pars intracerebral
PIP2	Phosphatidylinositol diphosphate
PK	Pyrokinin
PNS	Peripheral nervous system
POMC	Proopiomelanocortin
PPOMC	Pre-proopiomelanocortin
PRL	Prolactin-releasing peptide
PVG	Periventricular gray matter
PVK	Periviscerokinin
RER	Rough endoplasmic reticulum
SA	Short-axon
SNAP	Soluble SNF attachment proteins

SNARE	SNAP receptor protein
SNRI	Serotonin-norepinephrine reuptake inhibitor
STAI	State-trait anxiety inventory
T	Transporter
TC	Tritocerebrum
TCAs	Tricyclic antidepressants
TH	Tyrosine hydroxylase
TS	Tourette syndrome
VIP	Vasoactive intestinal polypeptide
VPL	Ventral posterolateral
VPM	Ventral posteromedial nucleus
VS	Ventral striatum
vT	Ventral thalamus

PREFACE

Neurochemistry is a burgeoning academic discipline that advances our knowledge of molecular, cellular, and medicinal neurobiology. Neurochemistry is a scientific discipline that investigates the role of chemicals in the formation of the nervous system, the function of neurons and glial cells in health and disease, the mechanisms of cell metabolism and neurotransmission, and the degenerative processes that occur in the nervous system. As a result, this book comprises chapters on a range of subjects divided into major categories: Intracellular Signaling in the Nervous System, Neural Processing and Intercellular Signaling in the Nervous System, Growth, Development, and Differentiation; and neurodegenerative diseases. The book contains in-depth reviews of each of these subjects. Neurodegeneration and neuronal disorders are widely discussed throughout the book and are a reoccurring issue in the majority of the chapters. This book will be invaluable to neurochemists and other scientists. Additionally, it will contribute to the education of present and future neurochemists and, ideally, will pave the way toward healing some of the most serious health problems.

While neurochemistry as a recognized discipline is a relatively recent development, the concept of neurochemistry dates all the way back to the 18th century. Originally, it was believed that the brain existed independently of the peripheral nervous system (PNS). Beginning in 1856, a series of studies contradicted the notion. The brain's chemical composition was almost identical to that of the peripheral nervous system. Johann Ludwig Wilhelm Thudichum, a pioneer in the subject of "brain chemistry," made the first significant advance in the study of neurochemistry. He was one of the first to propose that a variety of neurological disorders may be related to a chemical imbalance in the brain. He was also one of the first scientists to assume that the great majority of neurological illnesses might be treated, if not cured, using pharmacological techniques.

Neurochemistry became a recognized scientific study subject in the 1950s. The study of neurochemistry originated with a series of "International Neurochemical Symposia," the first of which was published in 1954 under the title Biochemistry of the Developing Nervous System. The International Society for Neurochemistry and the American Society for Neurochemistry were formed as a result of these sessions. These early conferences examined the putative neurotransmitter (NT) properties of chemicals such as acetylcholine (ACh), histamine, substance P, and serotonin. By 1972, concepts had become more solid.

The L-DOPA experiment was one of the first significant achievements in manipulating brain function using drugs. In 1961, Walter Burkmayer administered L-DOPA to a Parkinson's disease patient. Shortly after injection, the patient experienced a significant decrease in tremors and regained control of their muscles in ways they had not experienced in a long time. Within 2.5 hours, the impact peaked and lasted roughly 24 hours.

The book's content makes no presumptions about the reader's prior knowledge of neurochemistry or neurology beyond what would be taught in a typical secondary school. It is intended for students of neurochemistry who seek to get a better understanding of the relationships between the human brain and its functions, as well as the applications of neurochemistry in different areas.

CHAPTER

1

Introduction to Neurochemistry and Neural Functions

Contents

1.1. INTRODUCTION

The study of neurochemicals is the focus of neurochemistry, a discipline of neuroscience. An organic substance that contributes to neuronal activity is known as a neurochemical. This phrase is frequently utilized to refer to neurotransmitters (NTs) as well as other substances which alter neuron activity, like neuro-active medicines.

The roots of neurochemistry as a subject may be traced back to a sequence of "International Neurochemical Symposia," the 1st of which was titled *Biochemistry of the Developing Nervous System* and printed in 1954 (Brady et al., 2011). The *International Society for Neurochemistry* and the American Society for Neurochemistry was founded as a result of such sessions. The provisional nature of putative synaptic transmitter compounds like histamine, substance P, acetylcholine (ACh), and serotonin was debated at these initial sessions. By 1972, neurochemistry concepts had become more solid. Norepinephrine, dopamine (DA), and serotonin are examples of neurochemicals classed as "putative NTs in some neuronal circuits in the brain" (Figure 1.1).

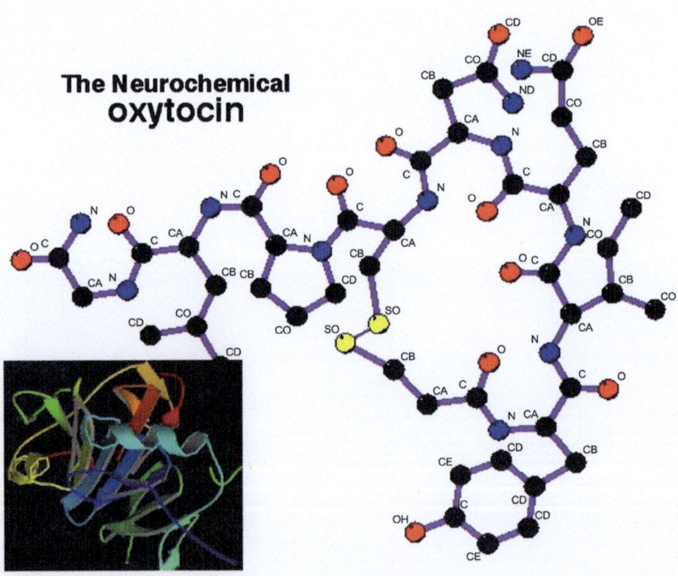

Figure 1.1. *Neuropeptide oxytocin.*

Source: https://www.researchgate.net/figure/Figure-Chemical-structure-of-oxytocin_fig1_228413599.

The following are some examples of neurochemicals:

- Oxytocin has a role in maternal behavior regulation. It is generated as a precursor protein inside magnocellular neurosecretory cells, which is then treated by proteolysis into its smaller active peptide state. Oxytocin is produced in certain areas of the brain, like the supraoptic nucleus, and works on cells in the ventral pallidum to cause the behavioral impacts of oxytocin. The brain produces a substantial quantity of oxytocin that is carried to the posterior lobe of the pituitary and released into the bloodstream, where it reaches target organs like the mammary glands (milk letdown). In the figure inset, oxytocin is represented coupled to neurophysin, a carrier protein.
- Despite being a gas, HNO_3 serves as a neurotransmitter. Since it is not liberated similarly to the other neurotransmitters, it is not grouped alongside them.
- The most frequent neurotransmitter is glutamate. GABA or Glutamate is secreted by the majority of neurons. Glutamate is stimulating, which means that it triggers nearby cells to fire action potentials when it is released by a single cell. "It's worth noting that glutamate is chemically similar to the MSG often utilized in Chinese cooking."
- Serotonin regulates emotion, sleep, as well as other aspects of life.
- A further instance of a neurotransmitter is dopamine. It is important for the limbic system that is involved in emotional function and regulation, to function properly.
- A neurotransmitter known as GABA is an inhibitory neurotransmitter.
- Acetylcholine is a neurotransmitter that helps with motor function (Figure 1.2).

Neurochemistry is a branch of science that studies the kinds, textures, and functioning of chemical elements present in the nervous setup (Heinbockel, 2018). These elements, in turn, control the neural system's physiology (Kandel et al., 2013). Little NTs, organic molecules, and neuropeptides are among the chemicals present in the nervous setup, and neurochemistry is involved with them. Neurological illnesses, such as Alzheimer's disease (AD) and Parkinson's disease, are frequently the result of abnormalities in the body's neurochemistry. Neurochemicals are used in medicine to change brain function and cure illnesses. Neurochemists look at how the elements of the nervous setup interact via procedures including learning, neural plasticity, development of the brain, and formation of memory, as well as how such elements alter as illness, brain malfunction, and aging progress. The chemical constituents of the nervous system would be introduced in this part, along with a detailed discussion of how internally and externally influence and affect such constituents (Bear et al., 2009).

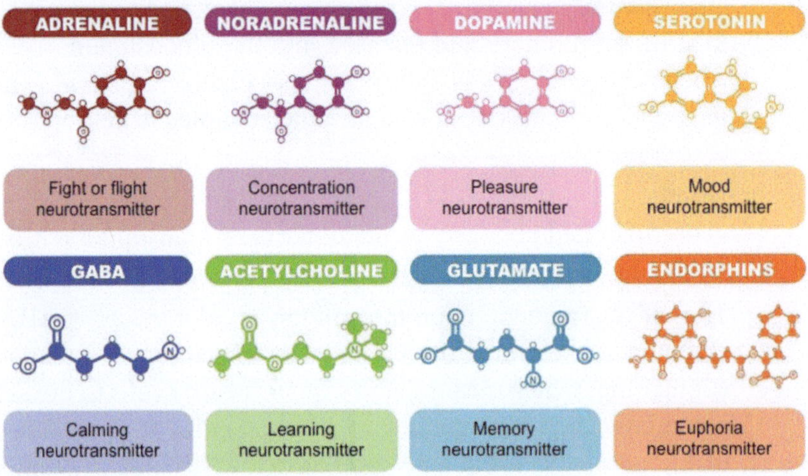

Figure 1.2. *Kinds of neurotransmitters.*

Source: https://ib.bioninja.com.au/options/option-a-neurobiology-and/a5-neuropharmacology/types-of-neurotransmitters.html.

1.2. BUILDING BLOCKS OF THE NERVOUS SYSTEM

The nervous system is made up of a large number of cells that differ in shape, function, as well as how they react with one another. Nerve cells, often known as neurons and glial cells are the two main kinds of cells. Both categories have several subclasses, each of which is called depending on its shape or function. Neurons are classed as interneurons, motor, or sensory depending on their role in neural networks, or as unipolar, bipolar, multipolar, or pseudo-unipolar depending upon their organization and existence of axons or dendrites. Axons transport data to other sections of the nervous system, whereas dendrites are the receiving component of a nerve cell. In brain circuits, although, this difference may be obscured since both dendrites and axons may perform either role. Adult humans have axons that exceed a length of one and a half meters, while bigger animals like giraffes have much longer axons. Axons transfer physical substance toward the axonal terminus and from the terminus to the cell body, as well as serve as long-distance communication mechanisms for data through the transmission of action potentials. Protein synthesis takes place in the cell body, which also houses genetic data. As a result, all proteins as well as organelles like mitochondria which are required in the axon terminus must be transported down the axon via motor proteins. Dynein and kinesin, two motor proteins in the axon, transport vesicles or organelles across the microtubules (Figure 1.3).

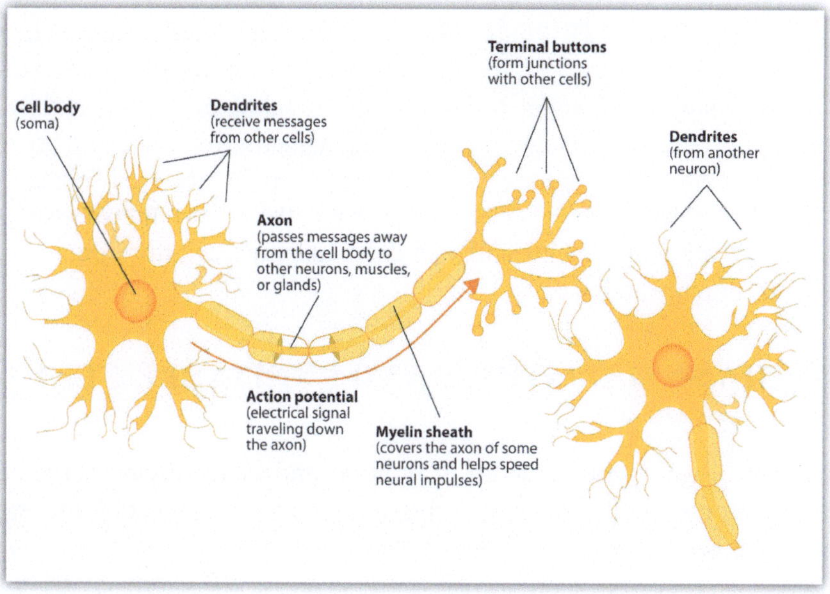

Figure 1.3. *The nervous system's structure and components.*

Source: https://openpress.usask.ca/introductiontopsychology/chapter/the-neuron-is-the-building-block-of-the-nervous-system/.

Kinesin transports vesicles from the endpoint to the cell body (movement of anterograde axonal), whereas dynein transports vesicles from the endpoint to the cell body (movement of retrograde axonal).

Glial cells are also available in a variety of flavors. Glial cells are becoming more well-known for their physiological roles in the nervous system, earning the moniker "true heroes of the nervous system" (Young, 2010). Myelin-forming Schwann cells and satellite cells are just discovered in the peripheral nervous system (PNS), whereas astrocytes (protoplasmic and fibrous), myelin-forming oligodendrocytes, and microglia are just discovered in the central nervous system (CNS). Although, these widely recognized categories, it must be obvious that nerve cells have such a wide range of morphology that it is nearly difficult to describe them accurately depending upon the shape, NT makeup, ultra-structure, localization, or physiology. Moreover, neurons with comparable structure and function are reported in far-flung animal species without current evolutionary ties.

Nerve cells have the mitochondria, smooth, and rougher endoplasmic reticulum, Golgi apparatus, nucleus, as well as other cellular components seen in many other cell kinds. The nucleolus and Nissl material in the cytoplasm are particularly intensely labeled when nerve cells are stained with dyes like cresyl violet toluidine blue (nucleic acid stain and Nissl staining) (Young and Baldwin, 2007). Bounded ribosomes on the ER Unbounded ribosomes in the cytoplasm are referred to as Nissl substances. This dark staining appearance reflects nerve cells' intense metabolism and constant creation of proteins and peptides, indicating that they are amongst the body's highly active cells.

1.3. CHANNELOPATHIES AND EXCITABLE CELL MEMBRANES

Membranes, which are made up of proteins and lipids, divide nerve cells as well as glial cells. Such proteins and lipids are essential for every neuron's specific functional involvement in the brain network, as well as the intracellular processes that take place in dendrites and axons far from the cell nucleus. Dendritic spine modification and axonal direction are molded throughout development concerning signal input at localized membrane sections that is transmitted to the interior of the cell via pathways and receptors. The interior and exterior of cellular membranes vary, allowing for an uneven work division due to an uneven division of proteins and lipids amongst axoplasmic and cytoplasmic leaflets (Xu et al., 2020). Lipids play a crucial role in the nervous system's function and structure. The membrane of lipids makes up the majority of the myelin that surrounds axons in both PNS and CNS. Membranes at the junctions, the linkages amongst nerve cells, contain distinct lipid contents. The connection has a synaptic apparatus comprising proteins and vesicles which assist in the specialized features of such membrane sections as well as synaptic plasticity (variation in synaptic propagation from pre-synaptic to postsynaptic neurons) (López et al., 2007). Lipid intermediates and lipid adjustment have a role in cell development signaling pathways, as well as influencing the function of trophic components and receptors (Heinbockel, 2017).

Nerve cells are stimulated cells with particular information-transmission characteristics. The membranes of the Nerve cell are designed with extremely critical pores or ion channels for Na, K, Ca, and Cl ions to do this. Such channels are essential for membrane stimulation and action potential transmission. Ion channels respond to voltage variations (channels of voltage-gated), chemical binding (channels of ligand-gated), and mechanical disruption.

Channelopathies, or ion channel diseases, are caused by impaired ion channel function as a result of abnormalities with subunits of the ion channel or having the control of the ion channels (Rose, 1998). When scientists discovered that mutations in a sodium channel subunit or a mutation in a gene that codes for a chloride channel in skeletal muscle resulted in hereditary muscle disorder, they called it "channelopathies."

The field of channelopathies has developed massively since then, with more than 100 publications in the last five years (Shiang et al., 1993). Mutations in a glycine receptor subunit or a nicotinic ACh receptor subunit have also been implicated in the development of channelopathies for ligand-gated ion channels (Steinlein et al., 1995). The main reasons for channelopathies may be found to be either hereditary reasons (such as congenital, arising from one or more mutations in the genes encoding the ion channel) or accumulated reasons (such as toxins, autoimmune invasion on an ion channel, etc.).

1.4. NEUROTRANSMITTERS (NTS) AND NEUROPEPTIDES

Neuropeptides and NTs are the most well-known neurochemicals because they can affect mind activity (Bear et al., 2009). One class of NTs is generated by ordinary amino acids like gamma-aminobutyric acid (GABA), glutamate, and glycine, which are all found in small amounts in the body. Such amino acids are involved in a variety of processes across the body. These are packed and kept in synaptic/secretory vesicles in the nerve endpoints of neurons so that they may be discharged via exocytosis in a calcium-dependent way when the neuron fires. After each synaptic discharge cycle, the vesicular membrane is regenerated or endocytosed, to prepare for the next one (Figure 1.4) (Argiolas and Melis, 1995; Werner and Covenas, 2010).

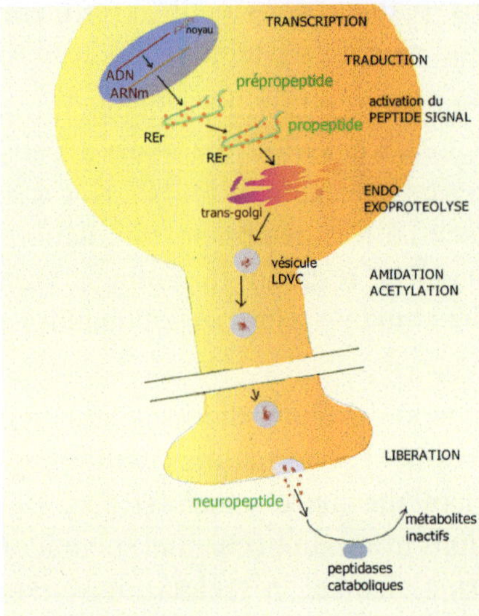

Figure 1.4. *Neuropeptide production.*

Source: https://www.differencebetween.com/difference-between-neuropeptides-and-vs-neurotransmitters/.

The very common excitable NT is glutamate. It is emitted at excitable synapses, causing membrane potential depolarization and perhaps excitatory firing in the postsynaptic cell. GABA, on the other hand, is the most well-known inhibitory neuro-transmitter. Neuronal excitation is reduced due to its effect. Glycine, an inhibitory NT present in the spinal cord, retina, and brain-stem is another inhibitory NT. Monoamines are a class of NTs entailed in certain types of memory, excitation, sensory processing, and sentiments regulation (Bloom, 1984; Jacob and Nienborg, 2018). Medicines are utilized to modulate their impacts in individuals with mental and neurological illnesses due to their functional responsibilities (Kurian et al., 2011). Monoamines have an amino group connected to an aromatic ring by a 2-carbon chain, as its name suggests. Monoamine oxidases are enzymes that stop monoamines from acting. Monoamines include epinephrine (adrenaline), DA, norepinephrine (noradrenaline), histamine, and serotonin. Because they include a catechol group, the first three are also referred to as catecholamines. NTs are discovered as residual amines like tryptamine, octopamine, phenethylamine, tyramine, as well as others. Opioid peptides, substance P, somatostatin, cocaine, and amphetamine-regulated transcript (CART), Oxytocin, orexin, neuropeptide S, glucagon, endorphin, dynorphin, neuropeptide Y, enkephalin, as well as other neuropeptides are examples of neuropeptides. Gaseous NTs include CO, nitrous oxide, and hydrogen sulfide (Burette et al., 2002). Such NTs are produced from scratch in nerve cells and can quickly permeate over the membrane of plasma to influence nearby cells due to their chemical makeup. To elicit skeletal muscle contraction, ACh is produced in the autonomic nervous system (ANS) as well as from motor neurons at the neuro-muscular endpoint. ACh is a chemical ester of choline and acetic acid. Since they are created from membrane lipids and are lipids, endogenously generated cannabinoids (endocannabinoids) like anandamide (AEA) vary from the previously stated neuro-transmitters. They may be quickly generated on request from the cell membrane and delivered non-synaptically rather than synaptically, as with traditional NTs (Heinbockel et al., 2016; Harvey and Heinbockel, 2018). Endocannabinoids control presynaptic NT discharge by binding to cannabinoid receptors (CB1R) on presynaptic neurons. -Endocannabinoids, along with gas NTs, are hence uncommon NTs (Barañano et al., 2001; Boehning and Snyder, 2003). The reality that all these new NTs serve as downstream mediators at synapses and pre-synaptically control whether GABAergic or glutamatergic synapses to modify discharge-probability in synaptic plasticity is one of their most notable features. Endo-cannabinoids and gaseous NTs are demonstrated to have a functional responsibility in experience-based activity and modulate a range of synaptic plasticity patterns (Cachope, 2012; Hardingham et al., 2013).

1.5. VARIABLES THAT HAVE AN IMPACT ON NEUROCHEMISTRY

What are certain of the variables that have an impact on the chemistry of the nervous system? Sensory inputs, ambient signals like recreational drugs, medications, as well as poisons, and physical changes like aging and disorder are all examples of variables that might alter neurochemistry and behavior (Tiraboschi et al., 2004; Simpson and Kelly, 2011). Instances of these variables that are recognized to have an impact on neurochemistry are outlined and described in further depth further down this page. Please keep in mind that this listing is not complete and that, in principle, practically any environmental stimuli or internal condition might have an impact on neurochemistry (Castillo et al., 2012; Katona and Freund, 2012).

1.5.1. Sleep

Circadian rhythms, which are a neurochemical (and potentially epigenetic) foundation, regulate sleep. Numerous brain areas are involved while waking, including the basal forebrain, lateral, and posterior hypothalamus, and tegmentum and pons nuclei. Monoamines and ACh, glutamate, and orexin or hypocretin are neuro-transmitters that have a role in wakefulness. The anterior or preoptic hypothalamic region, on the other hand, governs dynamic sleep processes, and peptide and GABA factors such as developing hormone-discharging hormone, prolactin, and cytokines improve sleep. Adenosine is a key homeostatic component that acts on A1 and A2A receptors in the pre-optic and basal forebrain regions. Sleep deprivation induces adenosine discharge and restoration of sleep via increasing inactivating nitric oxide synthase (NOS) in the basal forebrain. Several genes related to energy metabolism, neuronal propagation, synaptic plasticity, and stress defense are revealed to be differently regulated in sleep vs. wakefulness (Stenberg, 2007).

1.5.2. Exercise

A significant study has currently been conducted into the neurochemical variation that happens after and during exercise, with the findings that it provokes the upsurge of several chemicals, which includes neurotrophins, cortisol (CORT), lactate, such as BDNF, VEGF, and IGF-1, NTs such as norepinephrine, serotonin, DA, glutamate and ACh, and neuromodulators such as endocannabinoids Although, it must be highlighted that most of such modifications are proven peripherally, and there are still gaps in our understanding of wherein such variations happens in the brain. As a result, more research is required to establish a relationship between exercise-induced variations in peripheral levels and variations in central levels, as well as to determine how such substances are implicated in exercise-induced variations in cognitive, emotional, and other aspects of behavior (Basso and Suzuki, 2017).

1.5.3. Diet

The scientific discipline of "nutritional neuroscience" is associated with the impact of diverse diet components like micro-nutrients and macro-nutrients, including vitamins, minerals, food additives, and dietary supplements on neurochemistry. Nutrition is implicated in practically every facet of neurological function, influencing neurotrophic variables, neuronal networks, neurogenesis, and neuroplasticity, according to current studies on nutrition and its influence on the brain. This is unsurprising given that the brain utilizes a disproportionately big quantity of energy in comparison to the rest of the body. The human brain, for example, accounts for about 2% of total body mass but consumes about 25% of total energy input. As a result, systems included in the transport of energy from meals to neurons are expected to be critical for brain function regulation. Moreover, deficiencies in some vitamins as well as other co-factors, and the effects of metabolic diseases like diabetes, impact memory by modifying procedures in the body linked with energy management and the formulation of neuroendocrine and neurotrophic variables (that is, IGF-1 and BDNF) as well as NTs in neurons, which may impact synaptic plasticity, neurotransmission, and cell growth (Dauncey, 2009).

1.5.4. Stress

Stress is characterized by body-brain integration to an environmental or internal provocation that the body interprets as a disruption in homeostasis. The limbic system's reaction to stress includes both the activity of multiple NTs as well as the responsiveness of neurons to hormones and other chemicals produced by the adrenal cortex, namely glucocorticoids. As a result, the body-brain reaction is likely to play a significant contribution in the response to stress (Mora et al., 2012). Severe stress is linked to changes in NTs including ACh, DA, glutamate, and GABA in brain regions that control stress reactions. The amygdala, prefrontal cortex, hippocampus, and nucleus accumbens are among such regions. Glucocorticoids also have a function in certain parts of the brain, interacting with various NTs. Neuromodulators secreted from peripheral organs like the gonads (estrogens), pancreas (insulin), and liver (IGF-1) play a function as well. A long-term rise in glucocorticoid levels caused by a tough lifestyle might accelerate neuronal harm that happens in the aforementioned regions of the brain as people age. Stress depletion, on the other hand, can have an anti-aging impact (Mora, 2013).

1.5.5. Meditation

Yoga and Meditation are two examples of stress-relieving techniques. Such strategies have lately gotten more curiosity as a result of a growing body of evidence demonstrating both indirect and direct effects. Depending upon previous research, it is shown that meditation modulates the levels of NTs, including DA, GABA, norepinephrine, and serotonin in a manner that helps to alleviate psychological problems like anxiety. Meditation can also work as a sort of preventative medicine by lowering baseline levels of stress NTs and hormones (Krishnakumar et al., 2015).

1.5.6. Alcohol

Several NTs in the nervous system are affected by alcohol. Its main function is to promote GABA emission, and it does so primarily through GABAA receptors, resulting in sedative impacts (Banerjee, 2014). It also blocks postsynaptic NMDA excitable glutamate receptors, which adds to the sedative effect. Furthermore, alcohol provides euphoric impacts, which are more closely linked to DA surges. DA's impacts are also suggested to have a role in alcohol desire and relapse. Alcohol also affects opioid receptors, which might result in the release of β-endorphins. Enhanced serotonin and reduced nicotinic ACh receptors are two other key impacts.

1.5.7. Recreational Drugs

Medications may interfere with neurochemicals' normal functioning, hinder how they should operate, or impair their propagation (Korpi et al., 2015). Pleasure is frequently boosted initially, while intellectual capacity and reason are generally reduced. Psychomotor stimulants such as amphetamines, methamphetamines, and cocaine affect an excess supply of NTs, primarily the norepinephrine and the monoamines DA, and can also inhibit them from being absorbed back, resulting in an unusually substantial quantity of norepinephrine and DA in synapses, activating the mesolimbic DA setup (Hill and Thomas, 2011; Antoniou and Tseng, 2002). Medicines such as ecstasy (3,4-methylenedioxymethamphetamine) disrupt serotonin propagation and transit through brain circuits in the same fashion. Other drugs, like opioids, marijuana, and heroin, behave as activating neurons, agonists, and interrupting the natural propagation and generation of NTs. They mimic endogenous brain chemicals as well as bind to receptors as activating neurons, agonists, interrupting the natural propagation and generation of NTs. The brain may be reconfigured by neuroplasticity as it seeks to sustain chemical homeostasis after recurrent drug misuse (Enevoldson, 2004).

1.5.8. Neurodegenerative Diseases and Aging

Neurobiology of aging research is starting to reveal the processes that underpin not just the physiology of brain aging, as well as the mechanisms that render people more prone to cognitive impairment and neurodegenerative illnesses (Hung et al., 2010; Walker et al., 2019). The impairment of neurotransmission in age-associated illnesses including Alzheimer's and Parkinson's disorder has sparked research into the neurochemistry of the aging human brain. Age-associated alterations in the cholinergic, serotonergic, and dopaminergic systems are the most consistently identified of all the NT setups investigated (Strong, 1998). The DA system, particularly, is prone to deterioration with age (Dreher et al., 2008). The link between these NTs and emotion, cognition, and motor function can play a role in age-associated behavioral variation and predisposition to age-associated disorders (Pereira et al., 2012; Hou et al., 2019). Furthermore, abnormalities in certain neurochemical systems and other pathophysiologic procedures can interact to cause age-associated neurodegenerative disorders (Figure 1.5) (Leshner and Koob, 1999).

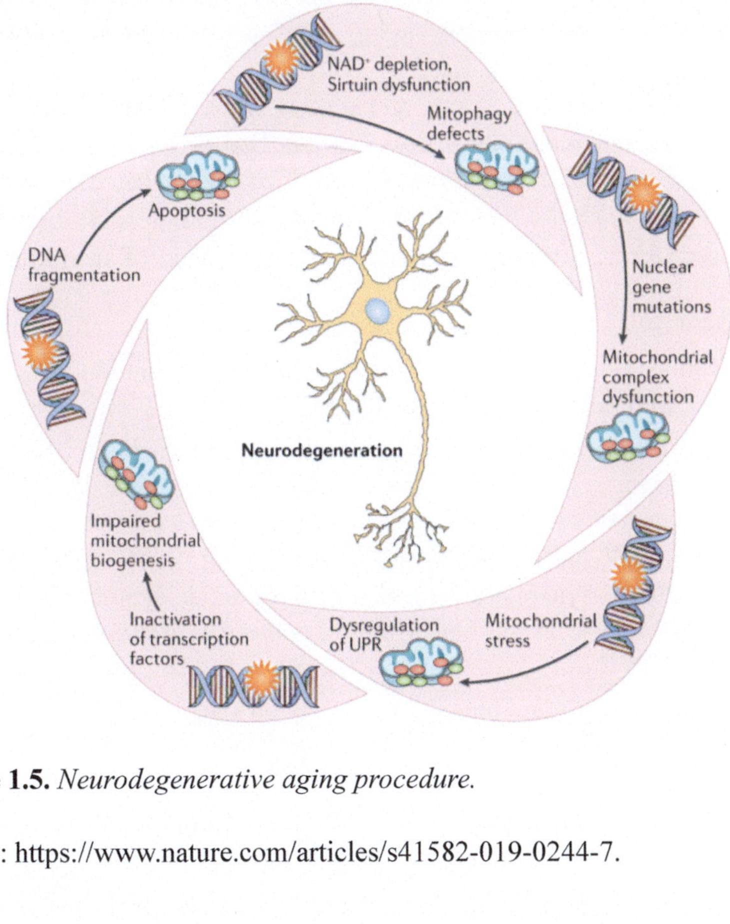

Figure 1.5. *Neurodegenerative aging procedure.*

Source: https://www.nature.com/articles/s41582-019-0244-7.

REFERENCES

1. Antoniou, T., & Tseng, A. L. I., (2002). Interactions between recreational drugs and antiretroviral agents. *Annals of Pharmacotherapy, 36*(10), 1598–1613.
2. Argiolas, A., & Melis, M. R., (1995). Neuromodulation of penile erection: An overview of the role of neurotransmitters and neuropeptides. *Progress in Neurobiology, 47*(4, 5), 235–255.
3. Banerjee, N., (2014). Neurotransmitters in alcoholism: A review of neurobiological and genetic studies. *Indian Journal of Human Genetics, 20*(1), 20.
4. Barañano, D. E., Ferris, C. D., & Snyder, S. H., (2001). Atypical neural messengers. *Trends in Neurosciences, 24*(2), 99–106.
5. Basso, J. C., & Suzuki, W. A., (2017). The effects of acute exercise on mood, cognition, neurophysiology, and neurochemical pathways: A review. *Brain Plasticity, 2*(2), 127–152.
6. Bear, M. F., Connors, B. W., & Paradiso, M. A., (2009). Neuroscience: Exploring the brain-2007. *Computational and Mathematical Methods in Medicine Gastroenterology Research and Practice Evidence-Based Complementary and Alternative Medicine, 2014* (Vol. 1, pp. 1–30).
7. Bloom, F. E., (1984). The functional significance of neurotransmitter diversity. *American Journal of Physiology-Cell Physiology, 246*(3), C184–C194.
8. Boehning, D., & Snyder, S. H., (2003). Novel neural modulators. *Annual Review of Neuroscience, 26*(1), 105–131.
9. Brady, S., Siegel, G., Albers, R. W., & Price, D., (2011). *Basic Neurochemistry: Principles of Molecular, Cellular, and Medical Neurobiology* (Vol. 1, pp. 1–19). Academic Press.
10. Burette, A., Zabel, U., Weinberg, R. J., Schmidt, H. H., & Valtschanoff, J. G., (2002). Synaptic localization of nitric oxide synthase and soluble guanylyl cyclase in the hippocampus. *Journal of Neuroscience, 22*(20), 8961–8970.
11. Cachope, R., (2012). Functional diversity on synaptic plasticity mediated by endocannabinoids. *Philosophical Transactions of the Royal Society B: Biological Sciences, 367*(1607), 3242–3253.
12. Castillo, P. E., Younts, T. J., Chávez, A. E., & Hashimotodani, Y., (2012). Endocannabinoid signaling and synaptic function. *Neuron, 76*(1), 70–81.
13. Dauncey, M. J., (2009). New insights into nutrition and cognitive neuroscience: Symposium on 'Early nutrition and later disease: Current

concepts, research and implications.' *Proceedings of the Nutrition Society, 68*(4), 408–415.

14. Dreher, J. C., Meyer-Lindenberg, A., Kohn, P., & Berman, K. F., (2008). Age-related changes in midbrain dopaminergic regulation of the human reward system. *Proceedings of the National Academy of Sciences, 105*(39), 15106–15111.
15. Enevoldson, T. P., (2004). Recreational drugs and their neurological consequences. *Journal of Neurology, Neurosurgery & Psychiatry, 75*(suppl 3), iii9–iii15.
16. Hardingham, N., Dachtler, J., & Fox, K., (2013). The role of nitric oxide in pre-synaptic plasticity and homeostasis. *Frontiers in Cellular Neuroscience, 7*, 190.
17. Harvey, J. D., & Heinbockel, T., (2018). Neuromodulation of synaptic transmission in the main olfactory bulb. *International Journal of Environmental Research and Public Health, 15*(10), 2194.
18. Heinbockel, T., (2017). Introductory chapter: Mechanisms and function of synaptic plasticity. In: *Synaptic Plasticity* (pp. 3–13). IntechOpen.
19. Heinbockel, T., (2018). Introductory chapter: Organization and function of sensory nervous systems. *Sensory Nervous System* (Vol. 1, pp. 1–35).
20. Heinbockel, T., Wang, Z. J., Brown, E. A., & Austin, P. T., (2016). Endocannabinoid signaling in neural circuits of the olfactory and limbic system. *Cannabinoids in Health and Disease*, 11–37.
21. Hill, S. L., & Thomas, S. H., (2011). Clinical toxicology of newer recreational drugs. *Clinical Toxicology, 49*(8), 705–719.
22. Hou, Y., Dan, X., Babbar, M., Wei, Y., Hasselbalch, S. G., Croteau, D. L., & Bohr, V. A., (2019). Aging as a risk factor for neurodegenerative disease. *Nature Reviews Neurology, 15*(10), 565–581.
23. Hung, C. W., Chen, Y. C., Hsieh, W. L., Chiou, S. H., & Kao, C. L., (2010). Aging and neurodegenerative diseases. *Ageing Research Reviews, 9*, S36–S46.
24. Jacob, S. N., & Nienborg, H., (2018). Monoaminergic neuromodulation of sensory processing. *Frontiers in Neural Circuits, 12*, 51.
25. Kandel, E. R., Schwartz, J. H., Jessell, T. M., & Siegelbaum, S. A., (2013). In: Hudspeth, A. J., (ed.), *Principles of Neural Science* (Vol. 1, pp. 1–17).
26. Katona, I., & Freund, T. F., (2012). Multiple functions of endocannabinoid signaling in the brain. *Annual Review of Neuroscience, 35*, 529–558.
27. Korpi, E. R., Den, H. B., Farooq, U., Vashchinkina, E., Rajkumar, R., Nutt, D. J., & Dawe, G. S., (2015). Mechanisms of action and persistent

neuroplasticity by drugs of abuse. *Pharmacological Reviews, 67*(4), 872–1004.

28. Krishnakumar, D., Hamblin, M. R., & Lakshmanan, S., (2015). Meditation and yoga can modulate brain mechanisms that affect behavior and anxiety-a modern scientific perspective. *Ancient Science, 2*(1), 13.
29. Kurian, M. A., Gissen, P., Smith, M., Heales, S. J., & Clayton, P. T., (2011). The monoamine neurotransmitter disorders: An expanding range of neurological syndromes. *The Lancet Neurology, 10*(8), 721–733.
30. Leshner, A. I., & Koob, G. F., (1999). Drugs of abuse and the brain. *Proceedings of the Association of American Physicians, 111*(2), 99–108.
31. López-Gil, X., Babot, Z., Amargós-Bosch, M., Suñol, C., Artigas, F., & Adell, A., (2007). Clozapine and haloperidol differently suppress the MK-801-increased glutamatergic and serotonergic transmission in the medial prefrontal cortex of the rat. *Neuropsychopharmacology, 32*(10), 2087–2097.
32. Mora, F., (2013). Successful brain aging: Plasticity, environmental enrichment, and lifestyle. *Dialogues in Clinical Neuroscience, 15*(1), 45.
33. Mora, F., Segovia, G., Del Arco, A., De Blas, M., & Garrido, P., (2012). Stress, neurotransmitters, corticosterone and body-brain integration. *Brain Research, 1476*, 71–85.
34. Pereira, M. D., Ksiazek, K., & Menezes, R., (2012). Oxidative stress in neurodegenerative diseases and aging. *Oxidative Medicine and Cellular Longevity* (Vol. 1, pp. 1–21).
35. Rose, M. R., (1998). Neurological channelopathies: Dysfunctional ion channels may cause many neurological diseases. *BMJ, 316*(7138), 1104, 1105.
36. Shiang, R., Ryan, S. G., Zhu, Y. Z., Hahn, A. F., O'Connell, P., & Wasmuth, J. J., (1993). Mutations in the α1 subunit of the inhibitory glycine receptor cause the dominant neurologic disorder, hyperekplexia. *Nature Genetics, 5*(4), 351–358.
37. Simpson, J., & Kelly, J. P., (2011). The impact of environmental enrichment in laboratory rats—Behavioral and neurochemical aspects. *Behavioral Brain Research, 222*(1), 246–264.
38. Steinlein, O. K., Mulley, J. C., Propping, P., Wallace, R. H., Phillips, H. A., Sutherland, G. R., & Berkovic, S. F., (1995). A missense mutation in the neuronal nicotinic acetylcholine receptor α4 subunit is associated with autosomal dominant nocturnal frontal lobe epilepsy. *Nature Genetics, 11*(2), 201–203.

39. Stenberg, D., (2007). Neuroanatomy and neurochemistry of sleep. *Cellular and Molecular Life Sciences, 64*(10), 1187–1204.
40. Strong, R., (1998). Neurochemical changes in the aging human brain: Implications for behavioral impairment and neurodegenerative disease. *Geriatrics (Basel, Switzerland), 53*, S9–12.
41. Tiraboschi, P., Hansen, L. A., Masliah, E., Alford, M., Thal, L. J., & Corey-Bloom, J., (2004). Impact of APOE genotype on neuropathologic and neurochemical markers of Alzheimer disease. *Neurology, 62*(11), 1977–1983.
42. Walker, L., McAleese, K. E., Erskine, D., & Attems, J., (2019). Neurodegenerative diseases and aging. *Biochemistry and cell biology of aging: Part II Clinical Science,* 75–106.
43. Werner, F. M., & Covenas, R., (2010). Classical neurotransmitters and neuropeptides involved in major depression: A review. *International Journal of Neuroscience, 120*(7), 455–470.
44. Xu, X., Cui, L., Zhong, W., & Cai, Y., (2020). Autophagy-associated lncRNAs: Promising targets for neurological disease diagnosis and therapy. *Neural Plasticity, 1*, 1–25.
45. Young, J. K., & Baldwin, K. M., (2007). *Elsevier's Integrated Histology* (Vol. 1, pp. 1–21). Mosby.
46. Young, J. K., (2010). *Introduction to Cell Biology* (Vol. 1, pp. 1–15). World Scientific Publishing Company.

CHAPTER

2

Neurochemistry and Neurochemical Systems

Contents

2.1. INTRODUCTION

At the biological level, there seems to be an irreducible aspect affiliated including all things labeled emotion. The purpose of this chapter is to define the nature of that aspect. Particular neurochemical systems underpin both consistent subjective readings of perceptions of people's emotions and wishes or impact, as well as consistent readings of expressive presentations, in biological feelings. As neither subjective nor the display readouts are necessarily conscious; in actuality, they usually work beneath the radar of conscious experience. Nonetheless, they have a significant impact on behavior.

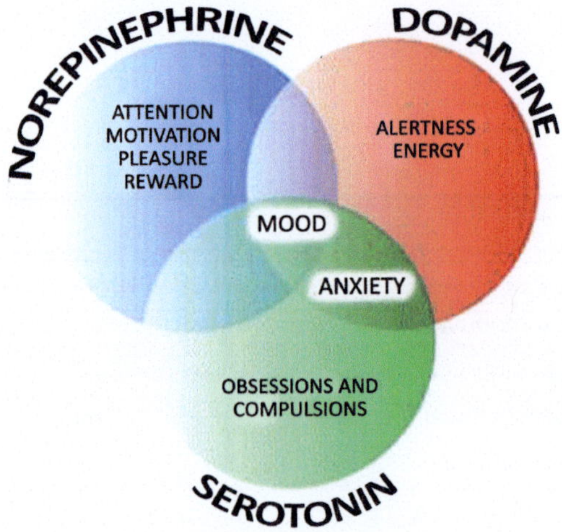

Figure 2.1. *Main kinds of neurochemicals.*

Source: http://www.sdkidsmentalhealth.org/neurochemicals-internal-medicine/.

The significance of neurochemical systems is explained in this chapter not because of their significance, but because of the proposal that such neurochemical systems are the objectively observable, irreducible, and manipulable basis of subjective efficacious qualia, that are encountered as wishes and emotions at the psychological level. Subjective experience is the most contentious and difficult facet of feelings from a scientific standpoint because it appears to be impossible to study objectively. Objective criteria, as well as manipulations of neurochemical systems (for example, via drugs as well as other sources, such as direct gene manipulation), are efficiently and reliably associated with emotional behaviors and displays in animals, as well as modifications in subjective effect described by humans (Figure 2.1) (Buck, 1993a, b; Panksepp, 1996, 1998).

Some neurochemicals are linked to particular subjective emotional qualia or tonalities, according to compelling and accumulating scientific evidence (Buck, 1993a). A particular feeling may be connected with a varied symphony/cocktail of distinct sensations and wants that may in theory be altered and assessed by manipulations and assessments of the underlying neurochemical systems. While not all neurochemical systems are associated with biological feelings, it might be argued that almost all biological feelings are associated with neurochemical systems. In general, such systems are certifiable, and their behaviors are manipulable and visible. Every given neurochemical might have a various impact on various people due to genetic or developmental/environmental impacts, as well as interactions with other systems, which may create individual variation in the neurochemical's impact (Figure 2.2).

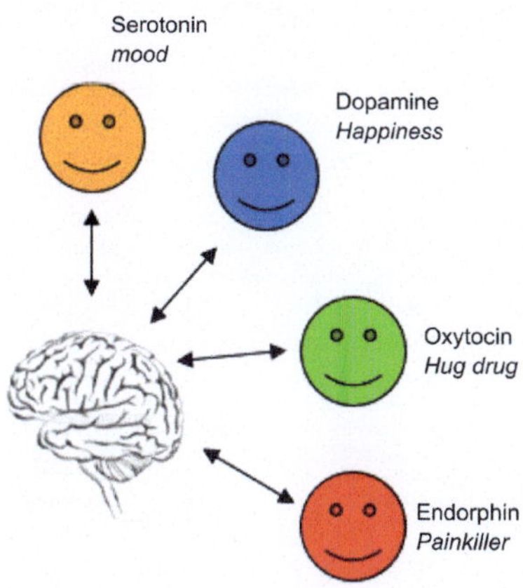

Figure 2.2. *Our brain produces happy hormones (neurochemicals).*

Source: https://www.thepositiveacademy.net/en/kurzy/master-your-brain-body/246/happy-neurochemicals.html.

2.2. NEUROMODULATION AND NEUROTRANSMISSION

2.2.1. Neurochemical Systems

Neurochemical systems are described as operationally ordered collections of neurons (specific nerve cells) that communicate with other cells, i.e., so both receive and transmit information from other cells, in the elaboration of Hebb's (1949) basic idea of "cell assemblies." In theory, and often, in reality, it is feasible to characterize biological feelings realistically in respect of particular neurochemical system activity. Biological feelings, in turn, underpin high-level intellectual, social, and ethical emotions: biological feelings are effective potential that arises during growth in an ecological framework. As a result, neurochemical systems are the inextricable core of all feelings.

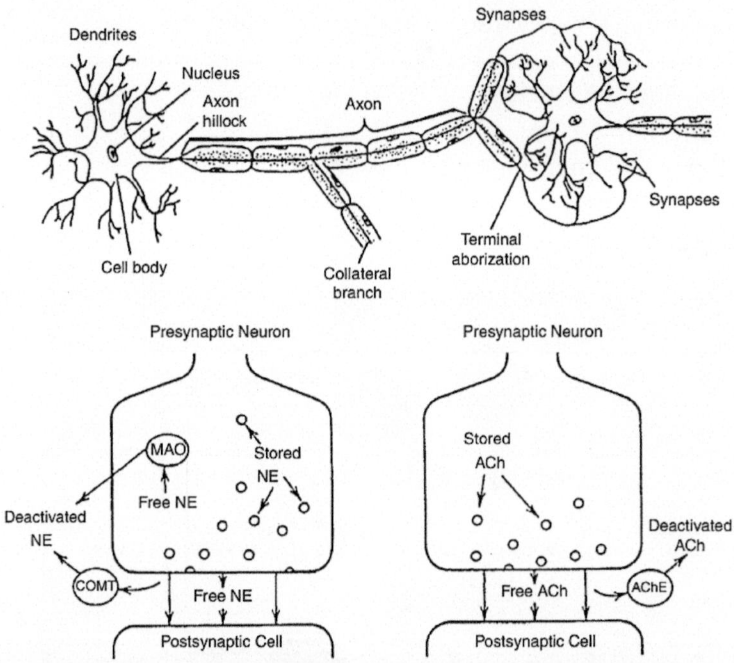

Figure 2.3. *Neurochemical systems. A simplified representation of a neuron (above), an adrenergic synaptic connection (left below), and a cholinergic synaptic connection (right below).*

Note: Norepinephrine (NE) is decomposed at the adrenergic synapse by monoamine oxidase (MAO) in the neuron and catechol-O-methyl transferase (COMT) in the synaptic gap. Acetylcholine (ACh) is decomposed in the synaptic space of cholinergic synapses by acetylcholinesterase (AChE).

Source: https://www.cambridge.org/core/books/abs/emotion/neurochemical-systems-evolution-and-function/6088CCD3F23FC234EE6F239D68F6AEAE.

Emotion is generally localized inside certain brain areas, according to studies of the brain and emotion. Although it may be beneficial, emotions may also be linked to neurochemical systems that span brain structures. Emotional events might also be physiologically dependent on visible and manipulable brain systems yet unrelated to every particular brain structure.

A neurochemical system's structural unit is the nerve cell (shown in Figure 2.3). A normal nerve cell or neuron communicates with other cells by transmission and reception of chemical signals via neurotransmitters (NTs), which are chemicals that convey signals along with tiny gaps among cells called synapses. The genes and several cells' metabolic machinery are found in the cell body of the neuron. Dendrites impact the neuron, causing it to become activated enough for a nerve impulse to be created. In a normal neuron, a nerve impulse is an electrical signal that travels down the axon. Chemical procedures are put in motion by nerve impulses at end buttons, structures at the axon's terminal, that lead to the release of NT molecules into the synaptic space. Consequently, the NT receptor sites in the membranes of the following cell in line, which might be dendrites of the next neuron or effector cells like muscle cells, exert their impact. The transmitting cell is known as a presynaptic cell, whereas the receiving cell is known as a postsynaptic cell (shown in Figure 2.3).

2.2.1.1. Neurotransmission

While the synaptic transmission is highly variable, certain common motifs may be recognized. To begin, NTs are generated inside the cell from precursor compounds, with certain enzymes influencing the process. On the final button, NTs are housed in vesicles. Secondly, NTs are discharged into the synaptic space via the presynaptic membrane, most frequently in reaction to the arrival of a nerve impulse moving down the axon. Thirdly, the transmitter material travels across the synaptic gap and attaches to receptor sites on the postsynaptic cell's membrane. The transmitter's chemical makeup fits into the receptor site similar to a key in a lock, activating it and enabling a passage of ions to flow through the membrane, briefly depolarizing the postsynaptic membrane and facilitating the postsynaptic cell's firing of a nerve impulse. Fourthly, the transmitter material is removed from the synaptic space, either through the action of an enzyme that breaks or disintegrates the NT or through re-uptake into the pre-synaptic cell, or even both (Pinel, 1993). Figure 2.4 depicts synaptic transmission for two important types of neurotransmissions: aminergic transmission including the norepinephrine (NE) and catecholamines dopamine (DA) as well as the indoleamine serotonin (or 5-hydroxytryptamine (5-HT)), and cholinergic transmission including the transmitter acetylcholine (ACh).

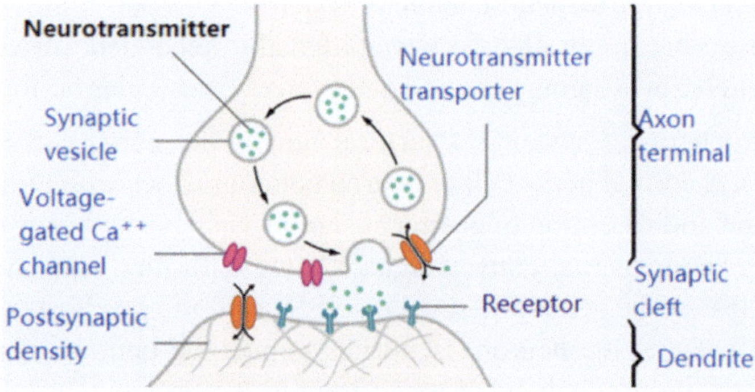

Figure 2.4. *The phenomena of neurotransmission as well as the function of neurotransmitters.*

Source: https://blogs.aalto.fi/neuralearn/2020/09/26/51/.

2.2.1.2. Neuromodulation: The Second Messenger

Neuromodulation is a very complicated procedure wherein the substance crossing the synaptic space is called a neuromodulator: it does not directly impact the postsynaptic membrane, but instead operates through a 2nd messenger, namely molecular structures inside the postsynaptic cell. In a nutshell, the neuromodulator functions as the main messenger, rather than directly modifying the postsynaptic cell membrane, it stimulates a G-protein linked to the receptor (Beatty, 1995). The 2nd messenger is the G-protein. It may start metabolic procedures that act as intracellular messengers, which may have a range of effects on the cell and the membrane. The membrane can be depolarized, stimulating the postsynaptic cell; alternatively, it can be hyperpolarized, inhibiting the cell, and rendering nerve impulse initiation very complex. The G-protein can also induce the manufacture of proteins inside the postsynaptic cell, modifying its inner metabolism and even influencing the expression of genes; impacting the synthesis of neurotransmission-related enzymes, and sprouting novel synaptic connections (DeWeid and Gispen, 1977). Memory and learning procedures may be influenced by these impacts. Such actions are frequently mediated via cyclic adenosine monophosphate (cAMP), a ubiquitous molecule with a broad range of physiological impacts (Matthews, 1998). The effects of the 2nd messenger may produce long-term or permanent changes in the cell's function, causing the cell to recall the impact. Neuromodulators are slow and endure longer than basic NTs, working on a time scale ranging from mins to hrs. and lasting weeks, months, or even years (Figure 2.5) (Grunberg and Straub, 1992; Purves et al., 1997).

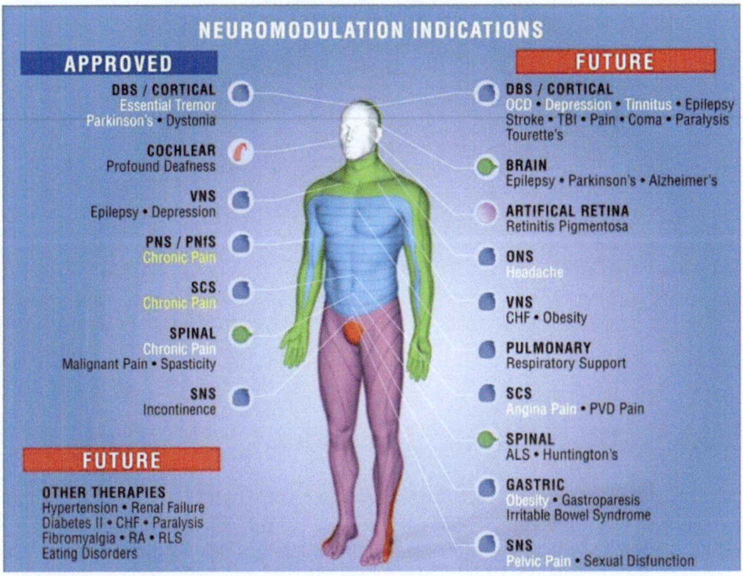

Figure 2.5. *Neuromodulatory instruments, including electrical and pharmacological, are used to treat a variety of human body diseases.*

Source: https://www.sciencedirect.com/science/article/pii/B9780123742483000021.

2.2.1.3. Receptor Sites

Based upon the receptor location with which a NT molecule binds, it might have distinct consequences. The transmitter molecule resembles an identifier into the receptor site, and since transmitter molecules have complicated 3-D structures, the similar molecule can fit into various kinds of receptor sites which are distributed separately under the brain's anatomy and can underpin various types of impacts that have implications for the sensitivity of drug and its adverse impacts. There has been a minimum of four types of DA receptor sites, denoted by the letters D1, D2, D3, and D4. Minimum 12 distinct receptor kinds are influenced by serotonin (5-HT). That enables a solitary kind of molecule to be accountable for complicated designs of actions that are often dissimilar and even contradictory (Figure 2.6) (Guastella and MacLeod, 2012; Guastella et al., 2012).

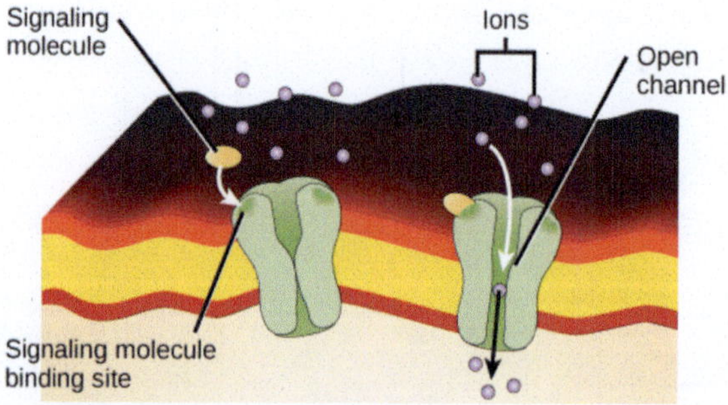

Figure 2.6. *Ion channel-associated receptors.*

Source: https://openoregon.pressbooks.pub/mhccmajorsbio/chapter/types-of-receptors/.

2.2.2. Genetic Biosynthesis of Peptides

Larger-molecule neurotransmission and smaller-molecule neurotransmission including peptides are the two major kinds of neurotransmission. As we will see, peptides are especially tightly related to subjective perceptions of moods and desires, and this association is especially strong. When peptides are formed, they start as relatively big precursor molecules that are the direct results of genes. A peptide-secreting neuron's cell body is where genetic transcription occurs, according to the cell's description. The transcription procedure leads to the formation of a pre-propeptide, which is then processed in the cell's endoplasmic reticulum, in which a segment of amino acids known as the signal sequence is excised from the peptide (Gur et al., 1995; Guerrero et al., 2005). The leftover material is a propeptide, which travels through the Golgi apparatus of the cell before being packed into vesicles for storage. In addition to physical transport, the vesicles undergo additional enzyme processing before being released at the end button. This procedure is referred to as fast axonal transport and happens as the vesicles are physically transmitted down microtubule paths in the axon from the cell body to the final button. Ultimately, the active peptide transmitters are discharged into the synaptic space, wherein they disperse and frequently impact many receptor sites close to one another. For example, pre-proopiomelanocortin (PPOMC) is a pre-propeptide, and proopiomelanocortin (POMC) is the equivalent propeptide that, in conjunction with particular enzymes, results in the production of the active peptides ACTH, MSH, and beta-endorphin, among other things (Figure 2.7) (Purves et al., 1997; Hahn et al., 2009).

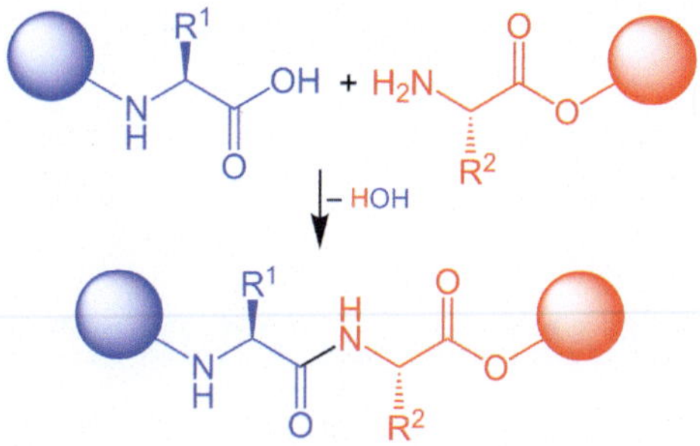

Figure 2.7. *Peptide production reaction.*

Source: https://en.wikipedia.org/wiki/Peptide_synthesis.

As neuromodulators and NTs, several peptides implicated in emotion have impacted both in the central nervous system (CNS), where they operate as neuromodulators and NTs and in the periphery of the body, where they work as hormones. It has long been established that peptide hormones produced by endocrine glands and released into the circulation perform many roles. Formerly, it had believed that peptide hormones could only be created by specific glands, such as the pancreas, the pituitary, and the adrenal glands (Hall, 1966; Haidt and Keltner, 1999). Peptides are now known to exist throughout the body, including the brain, the placenta, the vas deferens, and immune system cells, among other places. Many researchers believe that almost everybody cell can create virtually any peptide in modest quantities, and this is supported by research (Niall, 1982). A further point to note is that the central impacts of peptides on the brain are frequently associated with their peripheral impacts. In addition, the hormones oxytocin (OXY) and cholecystokinin (CCK) are implicated in the control of appetite and the regulation of childbirth, while corticotropin-releasing factor (CRH) is included in the production of ACTH as well as other stress hormones and the facilitation of avoidance behavior (Panksepp, 1993; Haidt, 2001).

2.2.3. Mechanisms of Drug Action

Drugs as well as other manipulations of the nervous system have the potential to alter the neurotransmission procedure in several ways. As a starting point, it is possible to either assist or hinder the production of the transmitter from precursors. It is also possible to either enhance or block the activity of enzymes that break down NTs in the brain. It is possible to enhance or block the re-uptake of transmitters into the presynaptic cell on a third level. A drug may also have the ability to act as a transmitter agonist, activating the receptor on the postsynaptic cell and thus simulating the effect of the transmitter; or it may have the ability to act as a transmitter antagonist, binding to the postsynaptic receptor site but not activating it, thereby preventing its activation by the transmitter (Figure 2.8) (Hess, 1928; Hertenstein et al., 2006).

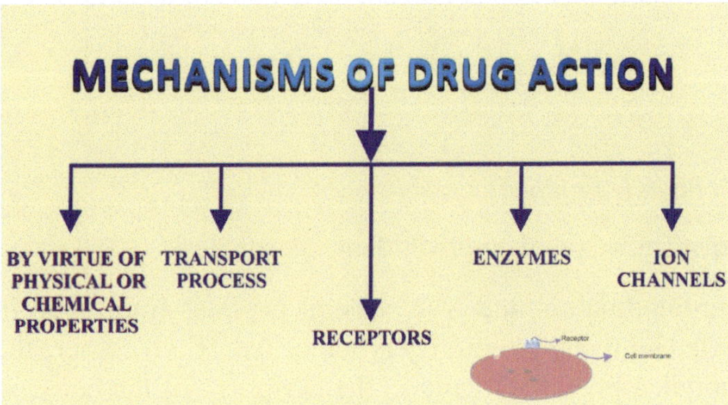

Figure 2.8. *Various mechanisms of drug action.*

Source: https://www.youtube.com/watch?v=tBEFtoH5XIc.

2.3. THE EVOLUTION OF NEUROCHEMICAL SYSTEMS

The essential chemical units of neural transmission originated long before the development of the spinal cord and vertebrate brain within the phylum Chordata, which is the class of animals that includes humans. The oldest evidence of the latter may be found in the fossilized remains of organisms dating back to the Cambrian period, which began 5,070 million years ago (Walker, 1987; Kohler, 2002). The classical transmitter molecules, on the other hand, have been around for a minimum of 1,000 million years and can be found in practically all phyla. They developed the capacity to activate a variety of ion channels in membranes, causing excitement, inhibition, or other impacts as a result of their action at membrane receptor sites. The homology among vertebrate and invertebrate receptors suggests that the fundamental receptor subunits originated a minimum of 800 million years ago in the presence of other organisms (Walker et al., 1996; Kolak and Volling, 2011).

2.3.1. The Evolutionary Antiquity of the Neurochemicals

When it comes to biological emotions, there are two noteworthy elements of neurochemical systems to consider. There are several reasons for this, one of which is the long evolutionary history of the chemicals that work in such systems, which developed long before the development of the vertebrate spinal cord and brain. Such molecules are not only old, but they are also primal. The configuration of the vertebrate spinal cord and brain evolved during the Cambrian era, along with all the other phyla of multi-celled animals that are currently found on the planet, such as arthropods, sponges, brachiopods, mollusks, chordates, and echinoderms as well as all the other phyla of multi-celled animals that are currently found on the planet (Dixon et al., 2001; Dixon, 2012). However, the segmented wormlike swimmer Pikaia, discovered among the Cambrian-era fossils in the Burgess shale, is the 1st known member of Chordata, the phylum that comprises humans and all other creatures with backbones (Figure 2.9) (Margules and Sagan, 1995; Koscik and Tranel, 2011).

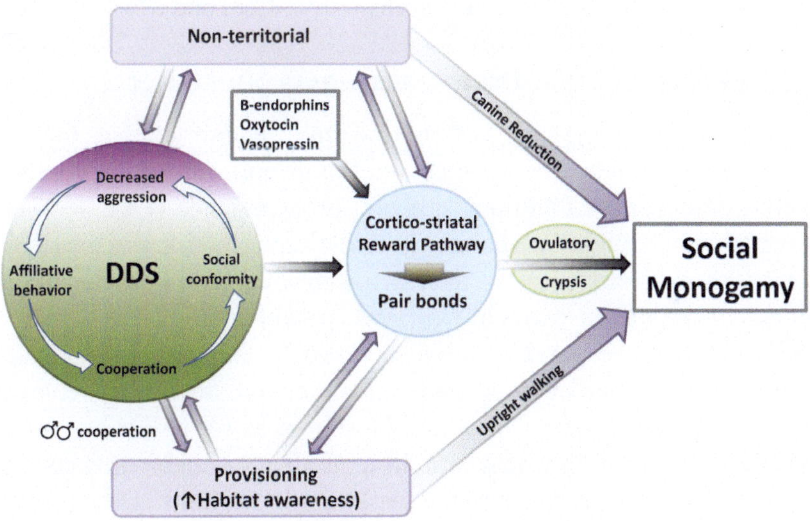

Figure 2.9. *Characteristics in the oldest known hominid neurochemicals.*

Source: https://www.pnas.org/content/115/6/E1108.

In contrary to the spinal cord/brain's relatively current history, finding demonstrates that several chemicals that today serve as NTs developed considerably earlier. Peptide neurohormones, for instance, are found not just in mammals' nervous systems, including humans,' as well as in insects, plants, segmental worms, and single-celled bacteria and yeast (Pert, 1985; Loumaye et al., 1982). Peptides found in single-celled protozoa are similar to those found in human adrenal-corticotropic hormone (ACTH), which is linked to stress; CCK, which is linked to satiety and digestion; the endogenous opiate beta-endorphin, which is linked to pain; insulin, which is linked to metabolism; and gonadotropin-releasing hormone (GnRH), which is linked to arouse sex and feelings (LeRoith et al., 1980, 1982; Miller et al., 1983). This means that such peptides should have presented in the shared ancestors of mammals, including humans, on the one hand, and single-celled yeast and bacteria on the other. These forefathers should have existed well over a billion years ago, long before the brain evolved. Furthermore, if peptides developed so initially in life's unicellular phase, the genes coding for such peptide receptors and molecules should have persisted and remained nearly unaltered throughout vertebrate and invertebrate history (Niall, 1982). Indeed, humans share roughly 40% of our genes with primitive single-celled living things: these genes should be employed in the development of receptors and peptides, which underpin the most basic forms of intercellular and intracellular communication (Landis, 1924; Lam and Marteleto, 2005).

2.3.2. Neurochemicals, Proto-Awareness, and Affects

The neurochemical systems' 2nd noteworthy feature is that they have been intimately linked to the most ineffable and enigmatic component of feelings: subjective emotional impact, or experience (Pert, 1997). The effects of medications that change the functioning of such neurochemical systems provide proof for all of this. Drugs may cause significant emotions in humans, like panic, extreme euphoria, peaceful or furious sentiments, decreased or increased anxiety and fear, sexual sensations, all in the lack of cognitive/rational justifications. Such human impacts are directly equivalent to animal impacts of a similar nature. However, as we will show, our fast-growing knowledge in this field contains complicated but broadly consistent data from a variety of resources (Langer et al., 1978; Lane et al., 1999).

Table 2.1 summarizes the impact of certain main medicines, including misuse substances, and is examined further in the perspective of certain NT systems. When looking at such pharmacological impacts, it becomes clear that there is a strong relationship between tangible and observable neurochemical systems on the one hand and ineffable and private emotional experience on the other. This is one method that public observation may be used to objectively operationalize and measure the emotional essence (Lindsley, 1951; Lindquist et al., 2012).

Table 2.1. Main Kinds of Psychoactive Drugs

Neurotransmitter System	Common Examples	Classification
Dopamine (DA)	Amphetamine, cocaine	DA re-uptake inhibitors
	L-DOPA	DA agonists
	Haloperidol, several antipsychotics	DA antagonists
	Amphetamine	DA releasers
Acetylcholine (ACh)	Nerve gases sarin and soman; nicotine Acetylcholine esterase; atropine; scopolamine	ACh agonists ACh antagonists
Serotonin (5-HT)	Psilocybin, LSD, mescaline	5-HT agonists
	Atypical antipsychotics	5-HT receptor antagonists
	SSRIs including fluoxetine (Prozac) TCAs including imipramine	5-HT re-uptake inhibitors
	Ephedrine	NE releasers
Endogenous opiates (endorphins)	Morphine, codeine, heroin, oxycodone	Agonists
	Naltrexone, naloxone,	Antagonists
Cannabinoid receptor	THC, cannabinol, cannabidiol	Agonists
Norepinephrine (NE)	Tricyclic antidepressants (TCAs)	NE re-uptake inhibitors
	MDMA (ecstasy)	5-HT releasers
N-Methyl-D-aspartic acid (NMDA)	Ethanol, PCP, nitrous, oxide ketamine,	NMDA receptor antagonists
Gamma-aminobutyric acid (GABA)	Ethanol, benzodiazepines, barbiturates, diazepam (Valium), chlordiazepoxide (Librium)	GABA receptor agonists

As Candace Pert described it in her book 'Emotional Molecules': We understood how the mind drugs opium, heroin, lithium, PCP, and valium entered the network and operated on receptors, as well as how endorphin, an endogenous or "in-house" molecule, communicated over a broad range. If we had to put a label on what such substances did, we would say they affected the emotional condition of the individual who took them, making her or him nervous, happy, sad, or calm. Whenever we concentrate on emotions, it becomes clear that the portions of the brain with the most receptors and peptides are also the parts of the brain that are linked to emotion expression (Pert, 1997).

It is important to note that subjectively reported impacts are associated with neurochemicals that have been there for thousands of years. CCK and Insulin, for instance, are peptides that allow the smallest single-celled animals to control their feeding behavior with their physiological requirements for a meal. The basic "proto-consciousness" of the need for meals that such organisms possess is evident in their behavior, which is characterized by the fact that they consume only when they are hungry. To be sure, this does not include the conscious feeling of hunger that more complicated organisms can be able to achieve. It is possible, however, that this primitive proto-awareness serves to create the biological and cognitive basis for the conscious subjective experience of emotion. Using sentiments and wants, such neurochemicals allow organisms to get access to and eventually grow into conscious awareness of, and their fundamental requirements, allowing them to survive and thrive (Martínez et al., 2004).

2.3.3. Voices of the Genes

As a result, old peptides linked to emotions and wants are direct products of certain genes. Peptidergic neurochemical networks allow such genes to interact, generating interoceptive perception systems that notify the organism of body conditions that are important for self-regulation. Effects serve as the genes' voices. Peptides are produced by genes and are linked to the subjective experience of sensations and wants. Genes do not rule the organism by their effects; instead, they cajole: entice, wheedle, convince, and coax. This gives the organism a variety of options for responding. Feelings of thirst and hunger, for instance, occur long before the body has a physiological need for drink or food, but they communicate that such requirements should be met with a tenacity that changes with the intensity of the sensation or desire. Fighting, escaping, making love, rejoicing, and sorrow are all identical. Impacts are constantly with us and just as we may direct our focused full attention to the feel of our shoes, we may always shift our focused full attention to our affective sensations and wants, but we normally do so only when they are at quite higher levels. The voices of the genes are constantly with us, although they generally speak in pianissimo. They do, however, yell and shout in unison at times.

The flexibility of reaction is enabled, if not required, by this structure. Biological and drives impacts, in contrary to comparatively simpler control systems like instincts and reflexes or set action patterns, do not regulate the organism's reaction. They define the agenda, specify the priorities, and prescribe the significance; however, the organism is in charge of the actual resolution. As a result, flexibility is the most inflexibly defining characteristic of human nature. As a result, free will is paradoxically biologically determined.

2.4. NEUROCHEMICAL SYSTEMS: STRUCTURE AND FUNCTION

We will look at some of the most common transmitters, such as GABA, ACh, norepinephrine (NE), DA, and serotonin (5-HT), and also peptides like the enkephalins. Certain chemical agonists may help NTs operate better, whereas antagonists may stop them from doing so (Pert, 1997). Several factors impact the responsiveness of a neurochemical system, such as the number and excitability of nerve cells in the system, the availability of NTs and their antagonists and agonists, anatomical location, and the level of sensitivity of receptor sites.

2.4.1. Small-Molecule Neurotransmitter (NT) Systems

Small-molecule NT systems and neuropeptides, which include comparatively larger transmitter molecules, are two types of neurotransmission systems (Purves et al., 1997).

2.4.1.1. The Transmission of Cholinergic

Otto Loewi, an Austrian physiologist, 1st established that neurotransmission involves chemicals in 1921, as well as the transmitter entailed, ACh, was identified in 1933. The muscarinic and nicotinic types of ACh receptors are named for their sensitivity to the harmful plant alkaloids nicotine (taken from the plant of tobacco) and muscarine (taken from a toxic red mushroom). The nicotinic receptor is quite simple: ACh is made in the nerve cell from choline and acetyl coenzyme A precursors in a procedure that requires the enzyme choline acetyltransferase (CAT). The formed ACh is kept in sacks at the presynaptic cell's terminal button, which is next to the synapse. Whenever an impulse travels down the axon, ACh is discharged into the synaptic space, in which it binds to ACh-sensitive receptor sites and temporarily changes or depolarizes the postsynaptic cell's membrane (Eccles, 1965). ACh molecules open a physiological gate in the postsynaptic membrane, enabling Na^+ ions and K^+ ions to pass via protein channels, causing the cell to depolarize. A potent enzyme in the synaptic region, acetylcholine esterase (AChE), quickly decomposes ACh, allowing the postsynaptic membrane to swiftly re-polarize (Purves et al., 1997). AChE decomposes ACh into choline and acetate, the remaining of which is re-uptake back into the presynaptic cell (shown in Figure 2.10).

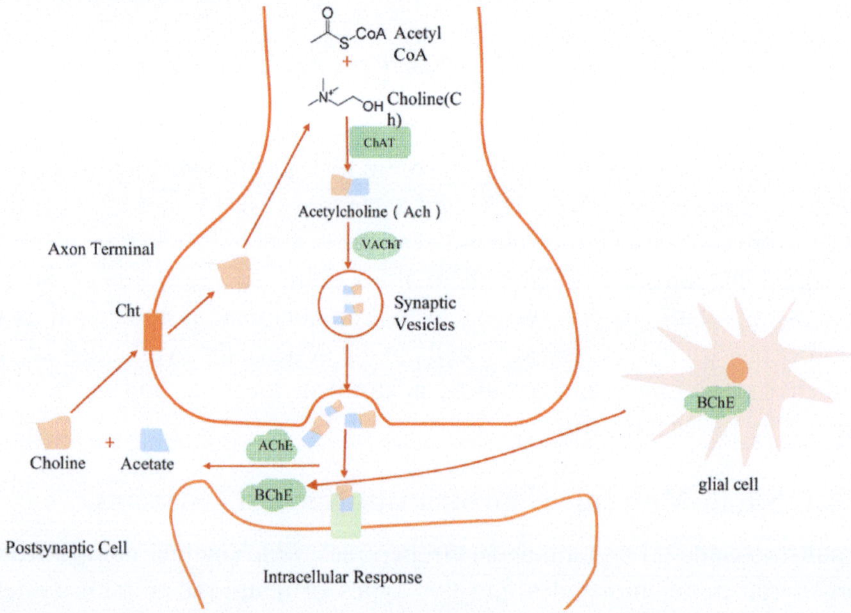

Figure 2.10. *The transmission of cholinergic.*

Source: https://www.researchgate.net/figure/Fig-3-Cholinergic-transmission-VAChT-neurotransmitter-transporter-ChATcholine_fig1_337748759.

Remember: Acetylcholine had been supplied into synaptic space as the [Ca^{2+}] ions in the axonal bulb rose, ChAT is equal to CAT; VAChT is equal to neurotransmitter transporter.

ACh is a peripheral NT found in the autonomic nervous system (ANS) and at the neuromuscular junction that causes skeletal muscular contractions by interacting with nicotinic receptors. It is also found in the CNS, however, its function there is not entirely clear (Purves et al., 1997). Muscarinic receptors, which include 2nd messengers and have more complicated and varied actions than nicotinic receptors, account for almost 99% of cholinergic receptors in the CNS (Beatty, 1995). Rather than immediately modifying the postsynaptic cell membrane, the ACh activates a G-protein linked to the receptor, which then serves as the 2nd messenger, causing a range of actions.

Numerous natural toxins target organ has the simple nicotinic ACh receptor, which acts to block off the receptor site, resulting in paralysis since skeletal muscles may no longer be stimulated by motor neurons., Snake venom, like that of the banded krait sea snake, and cobra, paralyzes prey in this way, as will the plant poison curare, which is utilized as an arrowhead toxin by people in southern America. Also utilized and misused as social drugs are drugs that operate on the nicotinic ACh receptor. The active substance in betel nuts and Tobacco, and are chewed in several regions worldwide (Purves et al., 1997). Each of these chemicals is extremely addictive, with addiction described as a high level of participation in and obsessive usage of a drug, as well as a high likelihood of relapse once the drug has been stopped (Jaffe, 1985). Nicotine is the most regularly utilized addictive chemical on the planet right now. The smoking of tobacco was brought from America to Europe, and nicotine was called after Jean Nicot, the French ambassador to Portugal in 1550, Jean Nicot was an early supporter of tobacco smoking's medical properties. Tobacco smoke is now widely recognized as being very hazardous to both the smoker and those in the surroundings, and highly addicting. Nicotine serves as a muscle relaxant on the periphery, although it has stimulating and arousing effects on the CNS, thanks to the release of norepinephrine and DA, as well as central cholinergic systems (Beatty, 1995).

Organophosphates, which include toxic and insecticides gases such as mustard gas, which was initially employed in the First World War, and the nerve gases Soman and Sarin, target cholinergic transmission. Sarin was a nerve agent utilized by Japanese extremists on the subway of Tokyo in 1995 and by Saddam Hussein, the Iraqi dictator, against Kurdish civilians in Iraq. Such chemicals block AChE, allowing ACh to accumulate in the synaptic space, causing muscle spasms, ultimate paralysis, and a protracted and agonizing death (Purves et al., 1997). Atropine (belladonna), a poison (deduced from deadly nightshade) that inhibits the muscarinic ACh receptor, counteracts sarin. Soldiers in modern militaries are armed with atropine, which they can use to inject themselves if they are exposed to nerve gas.

2.4.1.2. The Transmission of Aminergic

The transmission of Aminergic is mediated by biogenic amines, which are divided into two classes. The 1st class includes catecholamines, a group of related chemicals that comprise norepinephrine (NE), DA, and epinephrine (Epi). They are all related since they are all made from the same precursor, tyrosine. An enzyme inside the cell converts tyrosine to L-dopa, which is then transformed to DA, which is then transformed to norepinephrine, which is finally transformed to epinephrine. Histamine and indoleamine serotonin (5-HT) belong to the 2nd class of biogenic amines. The amino acid tryptophan is used to make serotonin. The significant ingrain in the case of DA, the locus coeruleus in the case of norepinephrine, and the raphe nuclei in the case of serotonin are all located in the brainstem nuclei (Figure 2.11).

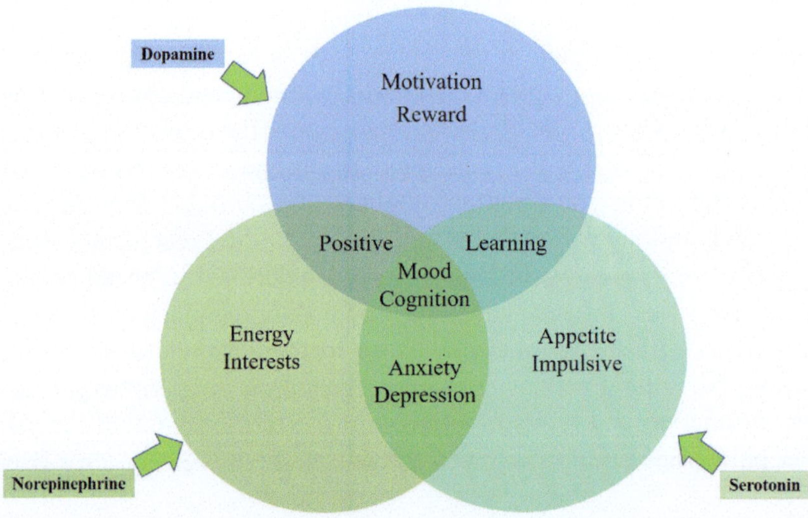

Figure 2.11. *Emotional roles of mono-aminergic neurotransmitters in main anxiety disorders and depressive disorder.*

Source: https://www.semanticscholar.org/paper/Emotional-Roles-of-Mono-Aminergic-Neurotransmitters-Liu-Zhao/31fd8925f38ca5f693d22751b4e6667723871dda/figure/0.

Figure 2.4 depicts the neurotransmission of amines. Neurotransmission of amines are produced at the presynaptic cell's end button and can be found both in sacks and in a free form. The enzyme monoamine oxidase deactivates them within the cell (MAO). Catechol-O-methyl-transferase inactivates them once they are transported into the synaptic area (COMT).

The mechanics of amine production and synapse function present various possibilities for interfering with aminergic activity. These are of special relevance since they are included in emotional procedures associated with a generalized elationde-pression continuum of behavior on a psychological level. The amines may be directly regulated by insertion or limitation of essential dietary ingredients like tyrosine, which is used to make DA and epinephrine. Sedation, weariness, and depression are common side effects of drugs that lower DA and epinephrine activity, like reserpine. To promote transmission of aminergic and hence operate like antidepressants, a range of medicines have been produced, and there are some of the most commonly prescribed pharmaceuticals in the globe. MAO inhibitors (MAOIs), like iproniazid, are antidepressants that work by lowering the concentration of amine-destroying enzymes in the cell, resulting in higher amounts of DA, epinephrine, and serotonin. Tricyclic antidepressants, like imipramine, adopt a different strategy, inhibiting amine reabsorption and so increasing their levels in the synaptic region via various methods. Specific Serotonin Reuptake Inhibitors (SSRIs), which function by specifically reducing serotonin reabsorption in the synaptic region, take a more discriminatory technique. SSRIs are antidepressants like sertraline (Zoloft), fluoxetine (Prozac), and paroxetine (Paxil) that have fewer adverse impacts than previous antidepressants (Purves et al., 1997). The epinephrine re-uptake inhibitor reboxetine has been studied for effectiveness and acceptability (e.g., Rampello et al., 2005), however selective epinephrine re-uptake inhibitors (SNRIs) are still not widely used.

Cocaine and amphetamine (Dexedrine), both of which have a long history of usage and misusage, are associated with aminergic neurotransmission on a psychological level (Wise and Bozarth, 1987). These are both DA agonists: amphetamine and cocaine block the reabsorption of DA while also increasing its secretion. Both contain antidepressant properties that improve emotion, increase attentiveness and energy, and decrease appetite: "The mind is full with a sense of purpose and energy as well as with strength" (Panksepp, 1981). The impacts of both drugs are comparable at large doses: expert cocaine addicts cannot tell the difference between pharmacologically equal injections of Dexedrine and cocaine (Fischman, 1984). Injections generate a pleasant and sexual peak that differs from that of opiate injections. Such euphoria is associated with feelings of increased mental and physical ability, the absence of a subjective requirement for meal or sleep, and reports of postponed and intensified orgasm in both males and females. With sustained usage, higher dosages are needed to attain the same degree of pleasure, and schizophrenia-like signs, such as hallucinations and paranoia, can develop (Beatty, 1995; Pinel, 1993).

2.4.1.3. Serotonin

The indoleamine neurotransmitter serotonin (5-HT) is engaged in a broad range of psycho-physiological processes. There is a minimum of 14 distinct types of serotonin receptors, which are grouped into seven separate families. A range of psychological diseases, such as anxiety, personality disorders, eating disorders, depression, drug-induced psychotic states, and schizophrenia, induce changes in such receptors (Roth, 1994).

Animals having lesions that disrupt serotonin action exhibit impulsive aggressiveness (Siever and Trestman, 1993), and analyzes of human research have shown connections among poor serotonin function and impulsivity, domestic violence, and anger, especially in males (Asberg, 1994; Brown et al., 1994; Coccaro, 1992; George et al., 2001; Tuinier et al., 1995). Failures to defer satisfaction or repress punishing behavior are common signs of such disorders, and New et al. (2002) found that low serotonin is linked to decreased activity in the orbitofrontal cortex (OFC), which is involved in planning and long-term behavior control (Figure 2.12).

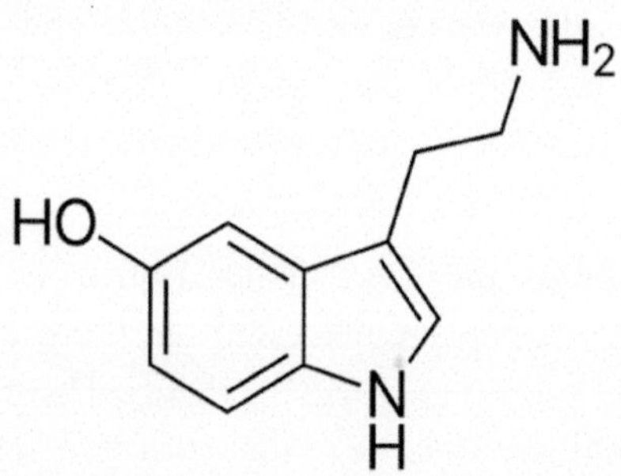

Figure 2.12. *Serotonin structure.*

Source: https://en.wikipedia.org/wiki/Serotonin.

Miller (1996) analyzed trials wherein serotonin levels had been experimentally boosted in normal specimens by Selective serotonin re-uptake inhibitors or administration of tryptophan, a precursor of serotonin, and found a connection between impulsive anger and low serotonin. Increased serotonin resulted in a decrease of aggression and an increase in collaboration, as well as an increase of confident dominance, or "social potency," as Carver and Miller put it. In a gaming environment, those with higher serotonin levels reported more dominating behaviors and were much more collaborative and extroverted, while also being assessed as low submissive and demonstrating more dominant behavior of eye-contact (Moskowitz et al., 2001; Tse and Bond, 2002a, b).

2.4.1.4. Monoamine Oxidase (MAO)

Monoamine oxidase, as the name implies, is an enzyme that breaks down the amines DA, serotonin, and norepinephrine. MAO-A breaks DA and eliminates norepinephrine and serotonin, whereas MAO-B breaks DA (Bortolato and Shih, 2011). Lower MAO levels result in increased amine levels, and natural MAO concentrations in people may be measured in the platelets of blood (Murphy, 1973). These concentrations reveal large individual differences, which are linked at the psychological level to humans and animals' behaviors that relate closely to a generalized extroversion-introversion component of personality. In rhesus monkeys, for instance, high levels of MAO are linked to an absence of social responsiveness, laziness, and passivity, whereas low levels have been linked to high play, activity, and social interaction (Redmond and Murphy, 1975; Redmond et al., 1979). Low MAO people see themselves as more sociable and have a stronger interest in social interactions in humans (Coursey et al., 1979).

Changes in the MAO genotype has been linked to a variety of human pathological situations, such as unsocial personality dysfunction (Bortolato and Shih, 2011; Reti et al., 2011), and there is an indication that such genetic potential correlates with earlier physical and sexual misusage. A high-activity MAO-A genotype plays a protective role man from initial maltreatment by making them less prone to develop unsocial issues later in life (Craig, 2005), whereas 30 years of longitudinal research found that men with the low-activity MAO-A variant who were revealed to be misuse as children were likely to report later hostility, offending, and conduct issues (Craig, 2005). Such findings suggest that a gene-environment association can account for part of the variation in developmental outcomes linked to physical and sexual abuse. Generally, MAO levels are inversely related to impulsiveness and sensation seeking and positively associated with inhibition (Castrogiovanni et al., 1990, 1991; Zuckerman, 1993; Zuckerman et al., 1980), and low and high MAO levels are linked to disinhibitory and inhibitory psychopathology, correspondingly (Castrogiovanni et al., 1990; Zuckerman, 1993; Zuckerman et al., 1980; Bongioanni, 1991; Sher et al., 1994).

2.4.1.5. GABAergic Transmission

GABA is another sort of small-molecule neurotransmission that is particularly significant to emotion. GABA is an inhibitory transmitter that prevents postsynaptic cells from depolarizing and firing. GABA is used as a NT by up to one-third of brain synapses (Purves et al., 1997).

The pyridoxal phosphate and enzymes glutamic acid decarboxylase (GAD), both extracted from vitamin B6, is used to make GABA from the precursor glutamate. Vitamin B6 deficiency has resulted in decreased GABA production, which has resulted in seizures due to the absence of inhibition. Active re-uptake into the presynaptic cell removes the majority of GABA. Barbiturates (e.g., phenobarbital), which are agonists or otherwise boost GABA efficacy, have been utilized as anti-seizure drugs and sedatives (Purves et al., 1997). Benzodiazepines (BZD) also seem to function by enhancing GABA's capacity to bind to its receptor, thus reducing neuronal activity. BZD include the anti-anxiety medications diazepam (Valium) and chlordiazepoxide (Librium). The amygdala, a brain region known to play a key role in scared and furious emotions, has a high density of BZD binding sites (Pinel, 1993). Anti-anxiety medicines are antagonistic to the peptide transmitter diazepam-binding-inhibitor (DBI), commonly referred to as the "anxiety peptide," as will be detailed later (Marx, 1985).

2.4.2. The Peptide Neurohormones

2.4.2.1. Definitions

Neuropeptides, as previously stated, include rather massive transmitter molecules (Purves et al., 1997). amino acids may be connected like beads on a thread by the carboxyl group of one amino acid as well as the amino group of another. The building blocks of peptides and proteins are amino acid chains. Peptides are two or more amino acid chains that are comparatively short (Pert, 1997). The form of the molecule, and hence the physiological activity of the peptide, is determined by the order wherein the amino acids are linked. Peptides, like other NTs, fit into receptor sites like a key in a lock, with the form of the key determined by the amino acid sequence. A peptide's impact may be dramatically altered by changing only one amino acid (Figure 2.13).

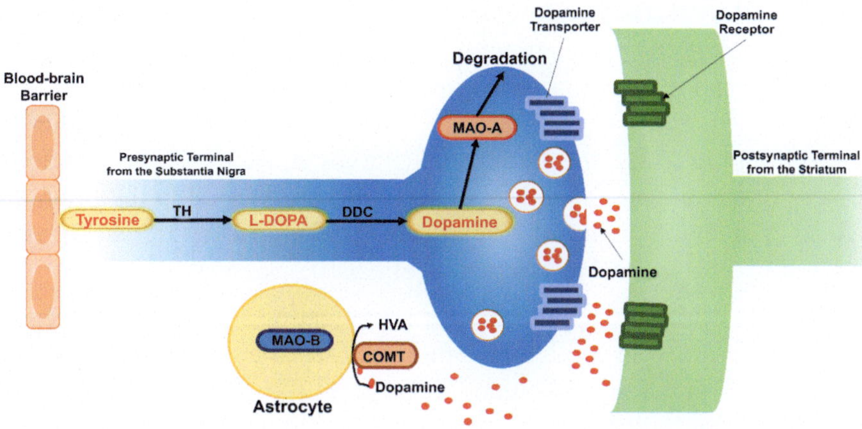

Figure 2.13. *The activity of monoamine oxidase (MAO) enzyme in neuronal cells.*

Source: https://www.researchgate.net/figure/Activity-of-monoamine-oxidase-MAO-enzyme-in-neuronal-cells_fig1_337876399.

Polypeptides are comparatively lengthy chains of roughly 100 amino acids or more, while proteins are polypeptide chains of more than 200 amino acids. Neuropeptides are short sequences of peptides that act as NTs and have comparatively quick but transitory actions. Neuropeptides outweigh small-molecule NTs by a large margin. Longer peptide chains, generally, act like hormones, with gradual but long-lasting impacts, which are generally induced by their diffusion through the bloodstream. Many of the neuropeptides were formerly thought to be hormones. The discovery of neuropeptides blurred the lines among hormones and NTs, resulting in the name peptide neurohormones, which encompasses both types of action.

Long and short peptides generally exhibit structural similarities as they share amino acid sequences: they likely have comparable forms, may fit into identical receptor sites, and can have similar impacts. The polypeptide beta lipotropin, for instance, is made up of 91 amino acid sequences. The 1st 41 of such sequences, according to Bolles and Fanselow (1982) describe adrenocorticotropin hormone (ACTH), which is found in single-celled animals and plays a significant role in the response to stress in humans. Melanocyte-stimulating hormone (MSH) is a pituitary hormone that stimulates melanin dispersion in pigment-consisting cells to darken the skin in reaction to UV radiation (tanning by the sun). The molecule beta-endorphin, a natural hormone produced by the pituitary gland, is made up of sequences 61 to 91. Beta-endorphin is a potent painkiller, as well as the name endorphin derives from a contraction of endogenous morphine. Methionine enkephalin (met-enkephalin) is a molecule made up of a segment of beta-endorphin sequences 61 to 65. Leucine enkephalin (leu-enkephalin) is identical but differs by one amino acid (Purves et al., 1997). In contrary to the hormone-like beta-endorphin, enkephalins are located in the brain and have less potent although faster-acting analgesic impacts: they can serve as NTs (Caldwell et al., 2010; Macbeth et al., 2010).

Recently developed approaches for investigating and modifying genetic systems have aided in the identification of particular functions played by neuropeptides in the regulation and influencing of behavioral responses. A more specific example is knock-out experimentations work is by breeding an animal wherein the particular gene that encodes a given peptide is removed; whereas knock-in experimentations work by breeding an animal having a gene transplanted on from some other species, which is the opposite of knock-out experimentations. Through targeting genes that create enzymes that are crucial to the transmitter's activity, such studies may also be used to examine small molecule transmitters, which are a type of NT. Because of the late 1980s, researchers have been working on developing the technology for these studies in mice (referred to as "transgenic mice"), which is now in widespread usage. The impact of knockout strains containing the peptides OXY and AVP, and serotonin and neuropeptides on social and emotional behaviors have been investigated (Haenisch et al., 2009; Lewejohann and al., 2010).

In non-vertebrates and vertebrates, genetic pathways for peptide production are comparable. The function of peptides as messengers is old and fundamental in evolution. The peptides predated the small-molecule NTs by a long time. The hydra belongs to the Cnidarians (which include sea anemones, corals, and jellyfish) and it has a nerve net that contains peptides but no serotonin, ACh, or catecholamines (Kandell et al., 2000), implying that the nerve system of the hydra's common ancestor, as well as other animals, functioned solely through peptide messengers.

2.4.2.2. Peptides and Subjectively Experienced Impacts

There are intriguing connections in between the functioning of old genes as well as the subjectively perceived sentiments and affective qualia, wishes with peptides. The peptides are dispersed in emotion-regulating regions of the brain, according to Pert (1985, 1997). She proposed whenever a peptide occupies a certain receptor site, this can bias the processing of information in a special manner, resulting in every peptide's "tone" producing a particular emotional state. Likewise, Panksepp (2003) claims that effective emotions are generated by neuropeptide-rich subcortical networks (Figure 2.14).

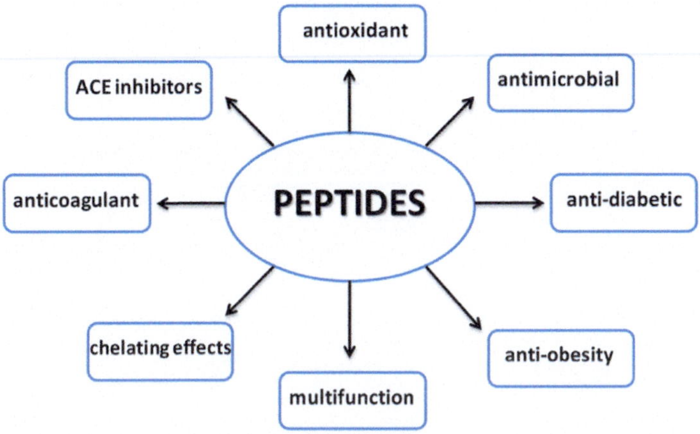

Figure 2.14. *Characteristics of bioactive peptides.*

Source: https://www.researchgate.net/figure/Properties-of-bioactive-peptides_fig2_342580697.

The next sections look at peptides that have been shown to have significant effective characteristics, starting with endorphins and endogenous opiates.

2.4.3. The Opiates

2.4.3.1. The Endogenous Opiates

Opium, which is derived from the opium poppy, has long been utilized and misused for its euphoric impacts and analgesic (pain-relieving) characteristics. By 380 BC, the Greeks had identified the euphoric properties of the wild poppy, and opium's main constituent, morphine, was called after Morpheus, the Greek god of dreams (Pinel, 1993). Pert and Snyder (1973) revealed that specific parts of the brain are susceptible to opiates such as heroin, opium, and morphine, as well as the opiate antagonist's naltrexone or naloxone, almost after 2,300 years. These parts of the brain are particularly linked to pain processes (Panksepp, 1991). Hughes et al. (1975) discovered opiate-like compounds in the mouse's vas deferens and subsequently in the brain. Over 20 opioid peptides have been discovered since then, divided into three categories: enkephalins, dynorphins, and endorphins. Pre-proenkephalin (proENK), p re-POMC, and pre-prodynorphin (proDYN), for example, are all linked to different genes or prepropeptide precursors (Purves et al., 1997). Opioids, like other peptides, are made directly by particular genes, unlike small-molecule transmitters, which are made from chemical precursors via enzymes (Figure 2.15).

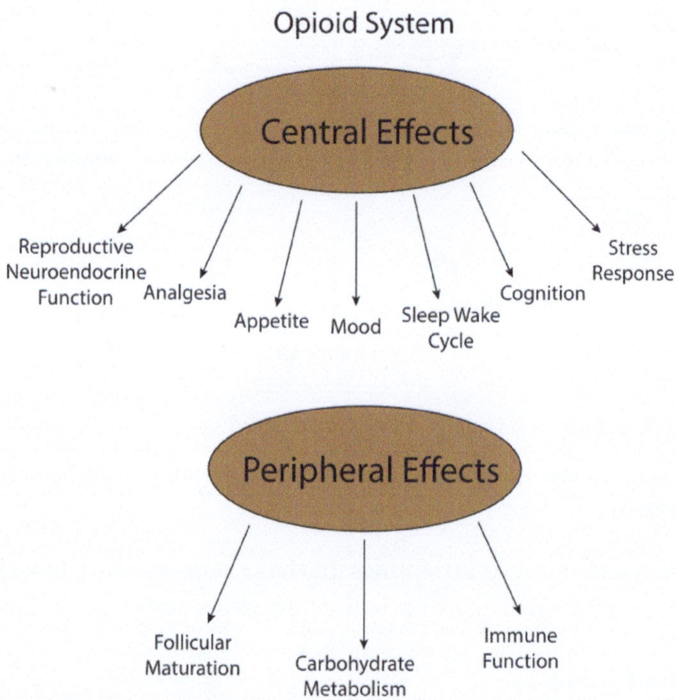

Figure 2.15. *Summary of endogenous opiate system.*

Source: https://www.chegg.com/learn/biology/anatomy-physiology-in-biology/endogenous-opiate.

2.4.3.2. The Pain and Opiates

Substance P is a pain-transmitting peptide NT found in the spinal cord. The more substance P is produced in these sensory nerves, the more pain is transmitted and perceived subjectively. Endorphins modulate the emission of substance P, which helps to control pain perception (Iversen, 1979). Opioid peptides limit the emission of substance P, resulting in the analgesic (pain-killing) actions of opiates. In many aspects, the resultant pain modulation system is comparable to Melzack and Wall's (1965) Gate Theory of pain control, which was hypothesized long before the role of opiates was discovered (Purves et al., 1997). Wall and Melzack postulated that brain impulses can activate gating circuits in the spinal cord, which may block incoming pain messages, explaining how emotional and cognitive variables might hinder pain perception.

Opioids work by essentially "shutting the gate" on pain in concert with other neurochemical systems. The analgesic impacts of opiates may be disrupted if serotonin is disrupted if particular serotonergic raphe nuclei in the brainstem are destroyed (Basbaum et al., 1976). Basbaum and Fields defined the mechanism of opiate analgesia as we know it today (Basbaum and Fields, 1978, 1979). The first stage is to stimulate opiate-sensitive neurons in the periaqueductal gray (PAG), the gray matter surrounding the cerebral aqueduct, in the midbrain. Opioid medications, electrical stimulation, or inputs from higher brain centers, presumably including cognitive and emotional cues, may all excite such neurons (Panksepp, 1998). Periaqueductal gray cells project downward to serotonergic cells in the medulla's raphe nuclei (especially the nucleus raphe Magnus), which then extend downward to cells in the dorsal horn of the spinal cord. The serotonergic raphe neurons synapse on endorphin-consisting neurons, which, when triggered, limit substance P emission and thus shut the gate, lowering the pain message (Figure 2.16) (Beatty, 1995).

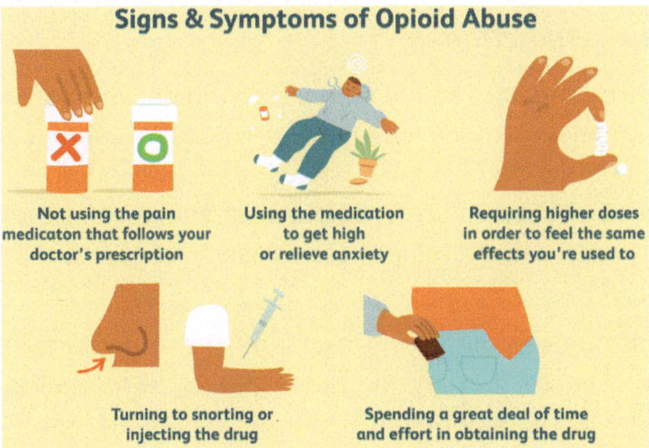

Figure 2.16. *Indications and signs of opioid abuse.*

Source: https://www.verywellmind.com/ten-most-addictive-pain-killers-22506.

The opiates' analgesic impacts may be prevented by the opiate antagonist naloxone that fights with the opiates for certain receptor sites and so blocks the opiates' impacts. The analgesic characteristics of both the placebo impact and acupuncture are dependent on endorphin pathways, according to studies utilizing naloxone (Purves et al., 1997). The placebo impact is described as a therapeutic reaction to a pharmacologically inert material, and it is widely known that pharmacologically inert drugs like saline treatments may dramatically alleviate pain. Initial research has shown that naloxone may counteract the analgesic benefits of placebo after wisdom tooth extraction (Levine et al., 1978): such that, when naloxone was administered instead of an inert drug, the placebo effect vanished and pain intensity did not decrease. There is also an indication that naloxone may counteract the analgesic effects of acupuncture (Mayer et al., 1977). Evidence suggests that acupuncture and placebos have actual pain-relieving benefits because of their capacity to trigger opiates (Ter Riet et al., 1998). Janssen and Arntz (2001) investigated pain sensitivity in novice parachute jumpers under real-life stress, finding that, according to the opioid-mediated analgesia interpretation, a placebo resulted in lesser ratings of pain sensitivity than naloxone. They also discovered a significant rise in plasma B-endorphin shortly after the leap, which was linked to anxiety and a loss of control during the jump. Non-opioid processes can also be implicated, as evidenced by reports of placebo impacts that were unresponsive to naloxone (Amanzio and Benedetti, 1999).

2.4.3.3. Opiate Addiction

The significance of opiates in the regulation of pain, according to Panksepp (1991), is correlated to their involvement in the development of social reliance and addiction. As previously stated, addiction is defined behaviorally in perspective of obsessive involvement with and usage of a substance, as well as a high proclivity to relapse after ceasing usage of the drug (Figure 2.17) (Jaffe, 1985).

Other definitions of addiction highlight the development of physical reliance on a drug, as seen by the development of tolerance while usage and withdrawal upon stopping to consume the drug in issue (Beatty, 1995). In a drug, tolerance is defined as a gradual decrease in sensitivity to a drug produced by continuous usage, such that a particular dose has gradually lower psychological and physiological impact, and consequently progressively higher doses are required to get optimal effectiveness. This might be because taking medicine can enable the endogenous resources of the relevant NT to be closed down, the sensitivity of receptor sites to be decreased, or both. A similar mechanism can underlie the withdrawal syndrome, which includes symptoms linked with the termination of drug usage. The normal functioning of the transmitter system can be altered as a result of drug usage.

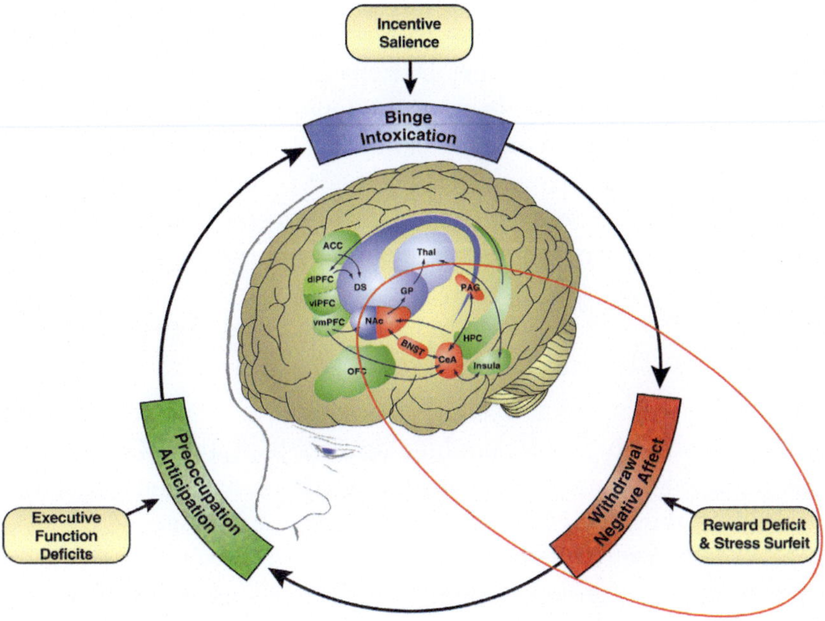

Figure 2.17. *Opioid addiction's neurobiology.*

Source: https://www.sciencedirect.com/science/article/abs/pii/S0006322319314350.

The addictive and euphoric impacts of opiates, like those of amphetamine and cocaine, can be connected to their impacts on DA reward systems: many addictive drugs, including benzene (alcohol) and nicotine, have a certain influence on the DA reward system (Wise and Bozarth, 1987). However, there are functionally significant variations between opiate-mediated reward (liking) and DA-mediated reward (wanting) that must be taken into consideration.

2.4.4. Other Neuropeptides

2.4.4.1. Diazepam-Binding Inhibitor (DBI): The Anxiety Peptide

Diazepam-binding inhibitor (DBI) is an 87-amino acid chain that was initially identified in the mouse brain in 1983 depending upon its capacity to displace diazepam (the anti-anxiety medicine Valium) from its GABA receptor binding sites as well as from the outer membrane of mitochondrial receptors

(Costa and Guidotti, 1991). It was later detected in the human brain as well as in the fruit fly *Drosophila melanogaster*, which is frequently employed in genetics studies because of its quick reproductive rate (Kolmer et al., 1994). The genes that encode Diazepam-binding inhibitors were sequenced and compared in rats and humans. Diazepam-binding inhibitors are linked to anxiety disorders, panic attacks, and depression in humans (Guarneri et al., 1990), and aggressive behaviors in rats (Marx, 1985; Kavaliers and Hirst, 1986). The diazepam-binding inhibitor is located in the amygdala, hypothalamus, discrete sections of the thalamus, cerebellum, neocortex, and hippocampus; as we will see, these are portions of the brain connected with scared and aggressive responses. Diazepam-binding inhibitors are detected peripherally in the testes, kidneys, adrenal gland, and liver (Josselyn et al., 1995; Fendt et al., 1995; Frankland et al., 1996).

Diazepam-binding inhibitor appears to have biological effects in addition to its role in regulating diazepam binding sites (Costa and Guidotti, 1991; Owens et al., 1989). Furthermore, the Diazepam-binding inhibitor molecule is similar to a previously investigated molecule termed acyl-CoA binding protein (ACBP) (Yanase et al., 2002). Notably, in light of its association with emotional activities, Diazepam-binding inhibitor appears to affect steroid synthesis in both the brain and the adrenal glands (Ferrarese et al., 1993).

Steroid hormones are lipid molecules that are membrane-permeable and diffusible, so they may enter and impact cells without functioning on particular receptor sites, unlike transmitters. Rather, they are frequently created in great quantities by specialized organs called endocrine glands and then circulated throughout the body between the circulations. The gonadal hormones are among the steroid hormones (the male androgens like testosterone generated by the testes; the female progesterone and estrogens generated by the ovaries). They also contain the adrenal cortex's corticoid hormones, like cortisol (CORT) which are commonly referred to as stress hormones. A kind of glial cell in the brain also produces these hormones, albeit in much lesser amounts. Glial cells, also known as neuroglia, are supporting cells in the CNS that have been revealed to perform crucial activities. The diazepam-binding inhibitor is found in steroid-allergic astroglia and Bergmann glia (glia that generates steroids in the brain) (neurosteroids). GABA receptor functions are regulated by neurosteroids (Figure 2.18) (Costa and Guidotti, 1991).

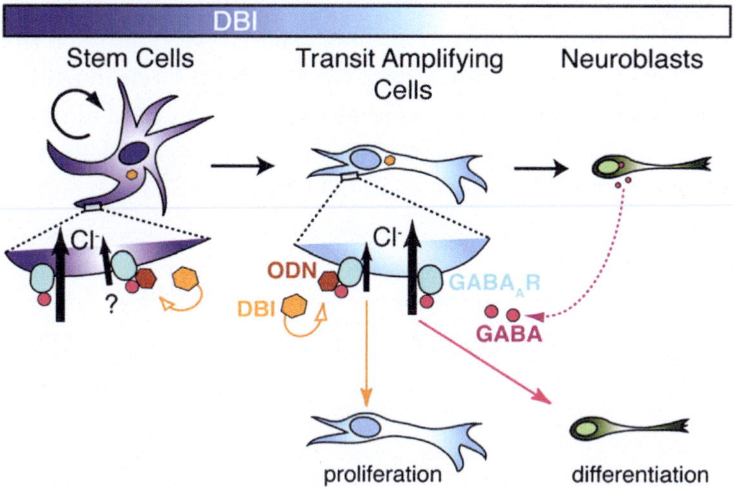

Figure 2.18. *Gamma-aminobutyric acid expression and impact in postnatal mice's subventricular zone.*

Source: https://www.sciencedirect.com/science/article/pii/S193459091100539X.

Another evidence that Diazepam-binding inhibitor is linked to stress is that Diazepam-binding inhibitor levels are highly connected with levels of corticotropin-releasing hormone (CRH) in the cerebrospinal fluid (CSF), indicating that Diazepam-binding inhibitor can have a role in the release of CRH in the human brain by inhibiting GABA (Roy et al., 1989; Roy, 1991). We will see how CRH causes the pituitary to create ACTH, which causes the adrenal cortex to release corticoid stress hormones (Ferrarese et al., 1993). The diazepam-binding inhibitor can be implicated in the activation and control of the hypothalamus-pituitary-adrenal (H-P-A) stress response, according to this data. Likewise, new research reveals that psychological stress generated by conditioned emotional cues enhances Diazepam-binding inhibitor expression of the gene in the mouse brain, but not physical stress (Katsura et al., 2002).

2.4.4.2. Cholecystokinin (CCK) and Panic

In unicellular organisms, CCK, a peptide implicated in appetite control, has been discovered (Loumaye et al., 1982). CCK promotes subjective sensations of anxiety and panic, according to a growing body of data (Bowers et al., 2012). CCK systems have been reported to influence freezing behavior and acoustic startle reactions in mice (Farook et al., 2001; Tsutsumi et al., 2001). A CCK antagonist was shown to diminish state anxiety in mice (measured by the animal's preference for a dark over a bright environment), but not trait anxiety (calculated via free exploratory test: Belzung et al., 1994). New predator and non-predatory fragrances were also reported to boost CCK gene expression in mice (Figure 2.19) (Hebb et al., 2002).

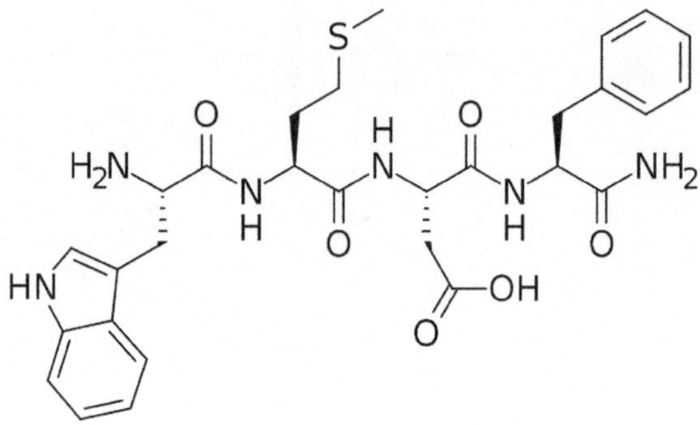

Figure 2.19. *Structure of cholecystokinin tetrapeptide (CCK-4).*

Source: https://en.wikipedia.org/wiki/CCK-4.

The CCK injections in people generate anxiety and terror without any external trigger (Abelson and Nesse, 1990; Bradwejn, 1993; de Montigny, 1989; Harro et al., 1993). Patients who suffer from panic attacks understand that the symptoms linked with CCK are similar to those they experience during their attacks (Bradwejn et al., 1991). Research of healthy volunteers found that a CCK antagonist significantly reduced the impact of a CCK agonist on subjectively reported apprehension, anxiety, and terror (Bradwejn et al., 1994, 1995). Healthy volunteers with preexisting extreme anxiety had more panicky worries and feelings of somatic indications, but they were no more susceptible to CCK-induced panic attacks than low or medium anxiety volunteers, according to Koszycki et al. (1993), and the authors stated that cognitive factors were not critical determinants of CCK-induced panic attacks. CCK also increases the flow of blood in limbic regions linked with fear, according to studies utilizing magnetic resonance imagery (MRI) and positron emission tomography (PET) scanning (Benkelfat et al., 1995; Javanmard et al., 1999; Reiman et al., 1989).

In people, there is a sign that CCK responsiveness is linked to personality (Kennedy et al., 1999). Self-administered anxiety measures, such as the state-trait anxiety inventory (STAI) and the Beck Depression Inventory, were completed by volunteers (Diazepam-binding inhibitor). The latency, quantity, severity, and duration of panic-like symptoms were then measured after they were given either a CCK agonist or a placebo injection. The STAI and Diazepam-binding inhibitor predicted CCK-induced symptoms, the fear reaction to panic symptoms, and post-injection anxiety, according to the findings (Jerabek et al., 1998).

Another research reveals that Gotu Kola, a traditional Chinese medication, may act as a CCK antagonist, lowering anxiety by binding to CCK receptor sites. Even though Gotu Kola did not affect self-reported mood, it did reduce the acoustic startle response in people (Bradwejn et al., 2000). Other data show that endogenous cannabinoid activation inhibits CCK release (Chhatwal et al., 2009; Javanmard et al., 1999), perhaps adding to tetrahydrocannabinol's soothing effects (THC: e.g., cannabis and marijuana). Bowers et al. (2012) found in a review that endocannabinoids have a function in the extinction of unpleasant learning.

2.4.4.3. Oxytocin (OXY) and Nurturance

Victor de Vigneaud produced OXY, the 1st peptide generated outside the body, in 1953. This was a defining moment in peptide study since it enabled the manufacturing of vast quantities of such compounds for both study and medication development. During labor, the pituitary releases OXY, which binds to receptors in the uterus and produces uterine contractions. Pitocin, a synthetic counterpart of OXY, is used to induce and accelerate labor. OXY has also been implicated in female sexual orgasm uterine contractions and lactation management (Pert, 1997). These OXY functions are unique to mammals, that are characterized by placental birth and mother's milk feeding of the young after birth (Gore, 2003). This behavior necessitates rather strong social relationships as well as bonding and nurturing traits that are absent in non-mammals.

OXY is significantly linked to maternal behavior in the CNS (Insel, 1992, 2002; Panksepp, 1992; Pederson et al., 1990). Many animals' OXY and maternal behaviors are influenced by sex steroids and offspring-related cues (Bale and Dorsa, 1995; Pederson, 1997). OXY is also linked to sexual behaviors. Increased OXY release is linked to both male and female orgasms (Richard et al., 1991). OXY increases the female sexual receptivity in rats, while OXY antagonists diminish receptivity (Arletti and Augusta Bertolini, 1992; Caldwell, 1992). OXY promotes sexual behavior in both males and females (Richard et al., 1991), lowers infanticide in animals (McCarthy, 1990), and reduces separation pain in animals (Richard et al., 1991;

Panksepp, 1992). Depending on whether a species is monogamous (mated for life) or non-monogamous, there are significant changes in the area of the brain related to OXY. Modern research examines this research, notably experiments using a little rodent known as the monogamous prairie vole (Donaldson and Young, 2008; Young, 2002). We will also discover that OXY is linked to human trust and social connection. OXY is a strong candidate for mediating sentiments of acceptance and social bonding, according to Panksepp (1993). In a double-blind trial in which OXY was delivered as a nasal spray, Rockliff et al. (2011) found that individuals who received OXY had improved their capacity to imagine compassionate feelings. Individual variations, on the other hand, impacted the effects of OXY: participants with high self-criticism and low self-reassurance, social safety, and attachment security had fewer positive reactions to OXY (Figure 2.20).

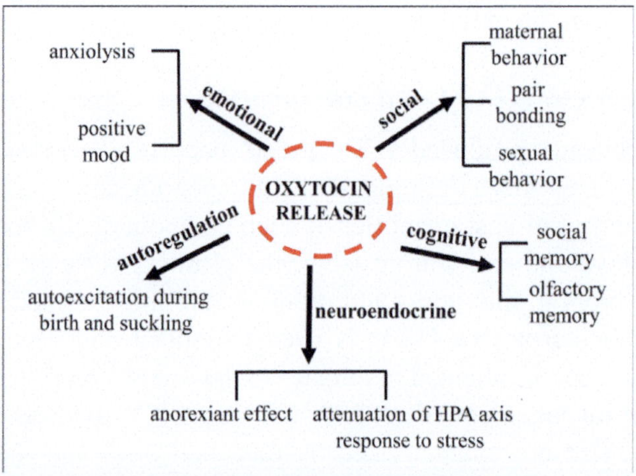

Figure 2.20. *Oxytocin regulatory control of complicated and various processes.*

Source: https://www.researchgate.net/figure/Oxytocin-regulatory-control-of-different-and-complex-processes_fig1_291518090.

Cardoso et al. (2012) showed substantial anxiety reduction with OXY only among women who were high in emotion-focused coping in another double-blind study employing intranasal delivery of OXY. Healthy volunteers were put through a ten-minute interaction with two other people (really confederates) who gradually began to neglect and finally fully exclude the participant. As we will see in subsequent chapters, such techniques are successful at isolating and ostracizing people. Only women with a high level of emotion-focused coping, a kind of coping linked to poor interpersonal functioning, were shown to benefit from OXY in reducing the anxiety caused by ostracism. That is ladies who said they reacted to stress with emotion and were thus prone to interpersonal stress experienced less anxiety after taking OXY. Men were less receptive to ostracism manipulation in general.

OXY does not exist as an active neuropeptide in non-mammals, which is one of the fascinating facts about it. OXY is generated from an evolutionary divergence of an ancestral hormone, vasotocin, which are detected in unicellular creatures (Loumaye et al., 1982) and enables socio-sexual reactions in reptiles and birds, together with arginine vasopressin (AVP) (Donaldson and Young, 2008; Moore, 1987). The basic activities of OXY and AVP are comparable in certain ways, especially when it comes to parental behavior, but they are very distinct in others (Carter et al., 2009). The fact that they share the same ancestral hormone raises issues about how analogous behaviors in reptiles, birds, and other non-mammals vary from those in mammals. OXY and AVP, in particular, have been linked to the divergence of female and male-typical sexual and aggressive-protective behavior patterns in animals.

2.4.4.4. Arginine Vasopressin (AVP)

AVP's peripheral actions include blocking water excretion in the kidney and high blood pressure. It was previously known as the antidiuretic hormone. AVP is linked to defensive behaviors, such as sympathetic nervous system responses, in the periphery (Figure 2.21) (Carter and de Vries, 1999).

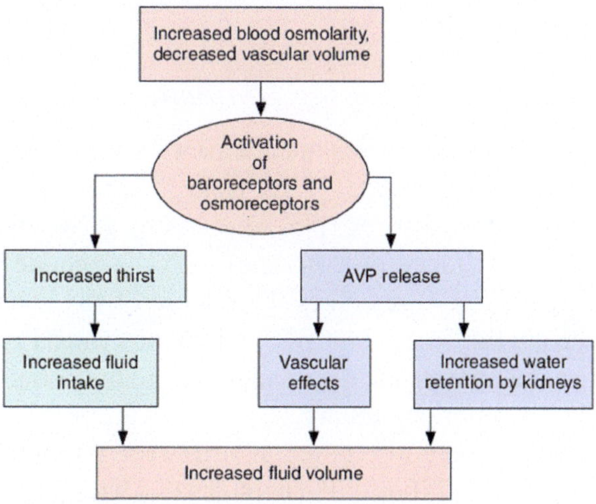

Figure 2.21. *Regulation of arginine vasopressin (AVP) release.*

Source: https://www.researchgate.net/figure/Regulation-of-arginine-vasopressin-AVP-release_fig1_5415629.

Although AVP has no effect on separation anxiety or women's sexual behavior in the CNS, it does increase male socio-sexual behaviors such as pair bonding in monogamous prairie voles (Windslow et al., 1993) and aggressive behavior (Koolhaas et al., 1990). AVP, in particular, causes selective intraspecies aggressiveness, possibly through the central amygdala (Gingrich et al., 1997). AVP antagonists, on the other hand, diminish male intraspecies aggressiveness in hamsters and rats (Potegal and Ferris, 1989; Ferris et al., 2008). In humans, nasal spray treatment of AVP increased aggressive inclinations (Cocaro et al., 1998), stimulated hostile facial displays in reaction to unknown men's faces (Thompson et al., 2006), and impaired identification of facial displays of negative emotions (Thompson et al., 2006; Uzefovsky et al., 2012). This result supports Panksepp's (1993) hypothesis that AVP is a specialized carrier for male dominance and persistence desires, controlling heightened male aggressive tendencies, and that the underlying subjective emotional correlate is irritation and anger.

AVP levels are linked to testosterone levels: androgens play a role in AVP production, and castration in rats lowers AVP neuronal networks in the brain by half (de Vries et al., 1985, 1994). In monogamous prairie vole males, it is AVP, not OXY, that supports mate preferences: AVP agonists increase the pair-bond formation and other monogamous behaviors, whereas AVP antagonists block them (Carter et al., 2008). Pitkow et al. (2001) found that transplanting the AVP receptor gene into the brains of male prairie voles enhanced AVP binding density, increased affiliative and anxiety behavior, and promoted pair-bonding.

2.4.4.5. Gonadotropin-Releasing Hormone (GnRH)

The hypothalamus, which controls the release of luteinizing hormone (LH) and follicle-stimulating hormone (FSH) from the anterior pituitary gland, regulates gonadal activity. The interplay of sex steroids (woman estrogen and progesterone; man, androgen testosterone) with the peptide gonadotropin-releasing hormone (GnRH) regulates the release of LH and FSH. Puberty begins when the peptide kisspeptin (KP) stimulates GnRH production, and GnRH is then produced from the hypothalamus into the portal system, which provides blood to the anterior pituitary. It stimulates the production of both LH and FSH in mammals and is also found in birds, reptiles, and fish (Schally et al., 1978). Indeed, GnRH is one of the peptides that have a long evolutionary history. The single-celled yeast Saccharomyces cerevisiae produces a peptide mating pheromone that is comparable to GnRH (Loumaye et al., 1982). It is fascinating, according to Loumaye and colleagues (1982), that a pheromone responsible for mating and zygote development in a unicellular organism also plays a major function in mammalian reproduction.

GnRH has several side effects, including boosting spermatogenesis in men and preparing the woman's body for ovulation by raising the synthesis of estrogen and progesterone in females and androgens in males. GnRH enhances sexual inclinations in the CNS; in fact, Panksepp (1993) argued that GnRH might be a major driver of human subjective sexual experiences, or libido (p. 94). Chemosensory and other sensory inputs in the brain of hamsters cause GnRH release, which aids mating behavior (Meredith and Fernandes-Fewell, 1994). Meredith and Fewell (2001) investigated the effects of electrical stimulation on the vomeronasal organ's chemosensory response. They used Fos, a neuronal activity marker, to map the brain's activation. The vomeronasal organ stimulated neurons in chemosensory pathways that were simultaneously engaged by mating and sexual chemical stimuli, and Fos was also activated in GnRH neurons in the brain's medial preoptic area. These GnRH neurons are thought to mediate chemosensory impacts on sexual behavior, according to the researchers (Figure 2.22).

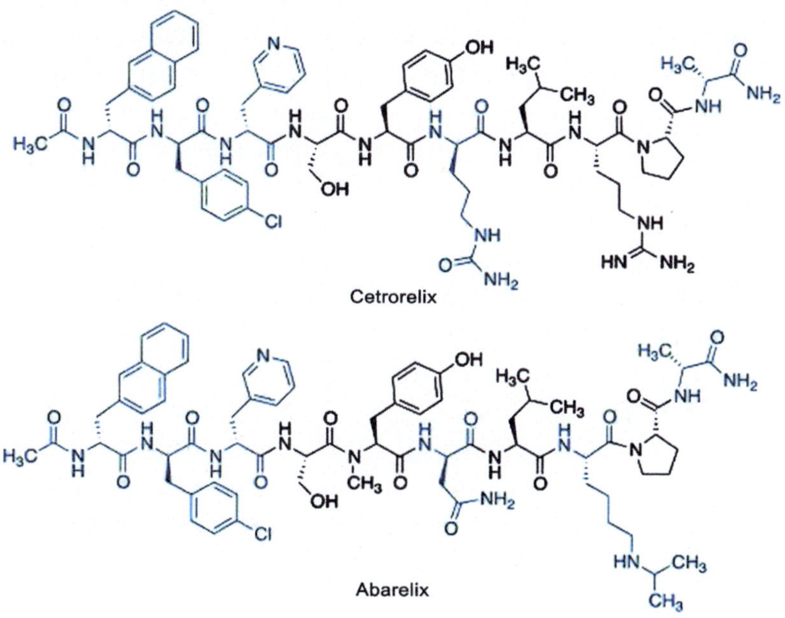

Figure 2.22. *Gonadotropin-releasing hormone (GnRH).*

Source: https://www.sciencedirect.com/topics/chemistry/gonadotropin-releasing-hormone.

It has been proven that GnRH exerts rewarding effects in male rats via their ability to induce conditioned location preferences in the male rats when introduced to a two-compartment chamber, rats are asked which compartment they prefer, and their responses are recorded. Once in one compartment, they are treated to medicine (in this case, GnRH), while in the other, they are subjected to saline placebo. After 24 hours, their choice is evaluated once more. A sensitive self-report assessment, as it were, of a drug's rewarding or punishing effects has been established: to the extent that a drug is rewarding, an animal would demonstrate a preference for the compartment that was linked with the drug. The rewarding effects of GnRH, on the other hand, are missing in animals that have been gonadectomized (de Beun et al., 1991). That the GnRH is rewarding because of its effects on gonadal steroids is inferred from the evidence.

2.4.4.6. Corticotropin-Releasing Hormone (CRH)

CRH is emitted from the periventricular hypothalamus and triggers the production of ACTH from the pituitary, which in turn promotes the release of several stress hormones, including steroids such as CORT, from the adrenal cortex. We know the link between CRH and DBI in more detail. It has been shown that CRH is connected with a coordinated central stress response in the brain and that this response has a significant influence on the development of anxiety/fear and separation distress in the brain (Panksepp, 1993). Alcohol, cocaine, opioids, and cannabis are all drugs of abuse, and the CRH has been implicated in mediating the unpleasant consequences associated with withdrawal from these substances (Rodriguez de Fonesca et al., 1997). It has been shown that stress experienced throughout childhood can generate long-term increases in CRH activity, as well as chronic dis-regulation of the hypothalamic-pituitary-adrenal axis response to stress (the H-P-A axis) (Figure 2.23) (Kolber et al., 2010).

A study conducted by Panksepp (1993) found that the CRH gene is connected with a specific type of chronic stress-induced depression or burnout in humans (Panksepp, 1993). Normally, the CRH shuts itself down in a negative feedback loop caused by increasing NE: CRH increases NE activity, while NE, in turn, suppresses CRH activity, resulting in a –ve feedback loop. It is possible that continuous stress-induced CRH activity may result in NE over-activation and, eventually, depletion of brain NE, which will be difficult to reverse at low levels of NE, and is no longer any mechanism to switch off the CRH (Risch, 1991). Consequently, persistent stress can cause semi-permanent brain alterations that, on a psychological level, may manifest as feelings of hopelessness and depression (Panksepp, 1993; Kolber et al., 2010).

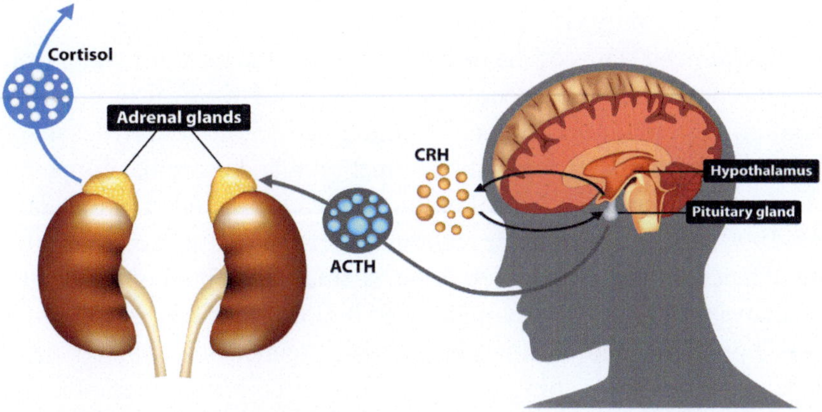

Figure 2.23. *Corticotropin-releasing hormone (CRH) in the pituitary gland.*

Source: https://www.news-medical.net/health/Corticotropin-Releasing-Hormone.aspx.

2.5. CONCLUSION

The evidence presented in this chapter shows that subjectively experienced psychological effects are linked to specific neurochemical systems and that the functioning of these neurochemical systems thus constitutes an irreducible essence associated with emotion. This review has demonstrated the diversity and complexity of neurochemical systems linked to subjectively felt effects, feelings, and desires, but it is also clear that converging evidence about how these systems work, derived from traditional animal research, new techniques involving gene manipulation, and evidence relating to drug effects in humans, is consistent, coherent, and expanding. To summarize, Table 2.2 lists the neurochemical systems we have looked at, as well as the types of subjective effects they have been linked to and the commonly used and abused drugs that are agonists or antagonists of these systems.

Considering the neurochemical underpinnings of biological emotions provides a conceptual platform from which other fundamental aspects of emotion may be explored further. A summary of some theorized links between the functioning of neurochemical systems and subjective experiences of sensations and wants, moods, and emotions is shown in Figure 2.24. I believe that the neurochemical systems connected with biological emotions serve as "readouts," or "voices of the genes," which have evolved to alert the organism to potentially critical information. Among this type of information are particular need states of the organism such as the need for water or food or sex; the need for heat or cooling; and the need for heat or coolness in association with main drives such as thirst, hunger, or sex. Arousal states, avoidance, and approach, agonistic competition, and prosocial cooperation, all of which are core functions of emotion, are also included in this data set (Buck, 1999). Every one of these readouts is active at all times: we are always aware of a certain level of hunger, thirst, sexiness, warmth/coolness; and we are also aware of a variable symphony of biological effects associated with arousal, approach, and avoidance, agonistic tendencies, and cooperative tendencies, among other things. Their presence in the brain is comparable to the presence of the feel of our shoes on our feet or the hardness of our chair, which are both constantly available for examination by the brain. Direct knowledge by acquaintance (KA) is the character of this consciousness, which was covered in the preceding section. However, we only pay attention to this information if it varies abruptly or if it is of a sufficiently high degree of importance. In this case, the level of arousal of the specific neurochemical systems responsible for the impacts is directly connected to the information provided.

I propose that in the absence of effective stimulation, neurochemical systems operate at a low level, similar to that of a pilot light and that they are generally accessed only when attention is intentionally attracted to them, as shown on the left side of Figure 2.24. As it were, the voices of the genes are whispering. For example, if the system is weakly stimulated by something as simple as changes in the day/night cycle or the pleasant scent of perfume in the air (S1), an individual can experience what most researchers regard as a "mood": it is relatively long-lasting, and the stimulus that caused it is not always readily obvious. In this case, the genes are murmuring and purring. Whenever the system is more efficiently stimulated, such as by a sudden loud noise, the sight of an attractive and receptive sexual partner, or the aggressive display of a rival (S2), the system is activated and appropriate readouts are organized, such as arousal reactions, displays, and subjective experiences that become dominant in the moment of stimulation. The genes suddenly scream out in this area. The arousal of what most psychologists define to be a "feeling" is, in my opinion, characterized by the fact that it is immediately experienced and easily related to the triggering event.

Table 2.2. General Drugs, Neurochemical Systems, and Their Effects

Drug	**Associated Subjective Experiences**	**Neurochemical Systems**
LSD, psilocybin, mescaline, MDMA (ecstasy)	Hallucinogen, euphoriant, paranoid horror (bad trip)	Serotonin (5-HT)
Cocaine, amphetamine	Wanting/pleasure; energy vs. sadness/ depression	Dopamine (DA) Norepinephrine (NE)
Morphine, heroin, codeine, oxycodone	Liking/elation, analgesia vs. pain sensitivity	Endorphins
Selective serotonin reuptake inhibitors (SSRIs), including fluoxetine (Prozac)	Confident dominance vs. depression (social)	Serotonin (5-HT)
Vasopressin (AVP)	Attraction, sexual arousal, protectiveness	Vasopressin (AVP)
Oxytocin (OXY)	In-group attachment, pair-bonding, trust vs. fear, suspicion, narcissism	Oxytocin (OXY)
Barbiturates	Depressant	GABA
Ethanol (alcohol), benzodiazepines including Librium and Valium	Anti-anxiety	Diazepam binding inhibitor (DBI)
Caffeine	Stimulant, focus	Adenosine antagonist
Nicotine	Stimulant	Acetylcholine (ACh)
Cholecystokinin (CCK)	Panic	CCK

Tetrahydrocannabinol (THC), cannabis, marijuana, hashish	Relaxation, sensory alterations	Cannabinoid
Gonadotropin-releasing hormone (GnRH)	Erotic feelings	Pituitary gonadal axis
Corticotropin-releasing hormone (CRH)	Stress vs. relaxation	Pituitary adrenal axis

When the stimulus is removed rapidly, the system quickly returns to normal (S2). However, if the neurochemical system has been overstimulated, including by a traumatic event like the 9/11 attacks on the United States, the system cannot be able to recover from its previous condition (S3). New neurochemical connections can have been formed, extra NTs generated, receptor sites expanded, and inhibitory neurochemical systems repressed, increasing the arousal and arousability of the intensely stimulated neurochemical system. The neurochemical system may now work at a level near to the threshold of awareness as a result of enhanced tonic arousal, making it reasonably easy to observe even with a modest stimulus (strong mood). Also, if the neurochemical system's arousability, or how easily it can be aroused, is improved, the emotion may be simpler to evoke in the future and fade more slowly (S4).

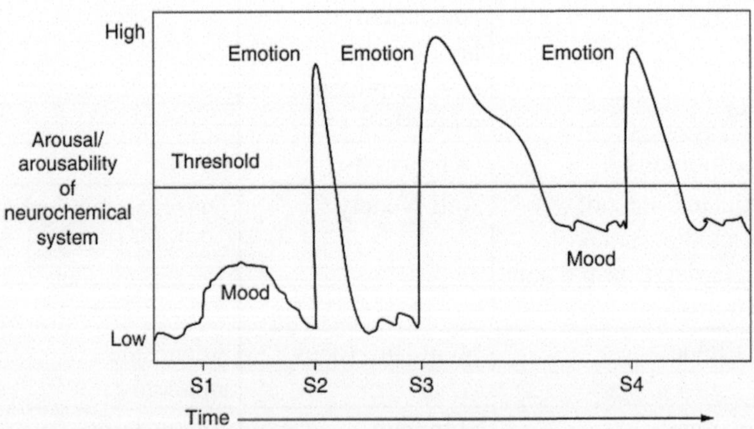

Figure 2.24. *Subjective experience and neurochemical arousal. Hypothesized relationships between the arousability and arousal of neurochemical systems as well as the subjectively experienced emotion and mood.*

Source: https://www.cambridge.org/core/books/abs/emotion/neurochemical-systems-evolution-and-function/6088CCD3F23FC234EE6F239D68F6AEAE.

The neurochemical systems that we may be regarded as semi-independent systems or modules capable of responding in the manner depicted in Figure 2.24, with each one adding to the overall symphony of perceived effect. According to this extension of the analogy, individual synapses may be compared to individual musical instruments, and neurochemical modules may be compared to groups of musical instruments, such as trumpets, clarinets, violins, and percussion instruments. More than that, various collectivities can be intertwined: for instance, the orchestral woodwinds, which comprise bassoons, clarinets, oboes, etc.; the orchestral brasses, which contain trombones, trumpets, and French horns, among others. These may be used in conjunction with one another, but not always. For instance, we discovered that the functioning of neurochemical systems including DBI and CRH frequently occurs in conjunction with other events. In addition, some elements, like percussion, can be active in practically all performances, whilst other sections, like strings, are more likely to have periods of stillness. Similar to sections of musical instruments, neurochemical systems that regulate arousal, reward, and punishment, agonistic competition, and prosocial cooperation can be compared to sections of musical instruments that regulate functions that are involved in virtually all emotions, such as arousal, reward, and punishment, agonistic competition, and prosocial cooperation. Synchronicity and asynchrony, disharmony, and harmony may occur in a symphony of desires and emotions, and these factors may have an impact on the entire performance. However, little is known about the factors that organize and direct the symphony of feelings and desires, although this is also something that may be discovered.

The nature of attachment, pleasure, love, and desire; the beginnings of despair; the emotional underpinnings of competition and cooperation; and the links between emotions and physical health can all benefit from this new perspective on biological emotions as founded on neurochemical systems. The concept of subjectively experienced impact as a symphony of arousal and arousability in various neurochemical systems provides an objective foundation for measuring, altering, and comprehending this most difficult component of emotion.

The functioning of neurochemical systems exemplifies and represents a common thread that runs through all emotional phenomena: dynamic regulation via communication. Communication is evident in neurochemical systems: a sender cell sends information to a receiving cell via chemical transmitter substances, and all biological emotions involve communication procedures. Bodily regulation, adapting the organism to environmental changes, and maintaining homeostasis are all important functions of such communication procedures. Furthermore, neurochemical systems are engaged in inter-organism communication, wherein information is expressed through displays in sending organisms and taken up by receiving organisms through pre-attunements to those displays (Buck and Van Lear, 2002). Social regulation, adapting a collectivity of organisms to environmental changes, and maintaining social organization are the primary functions of such communication procedures. All of these communication procedures, including the display mechanisms, the messages involved, and the receiving processes, may be objectively defined in the words of observable phenomena that are becoming more accessible to measurement. Ultimately, neurochemical systems are involved in interoceptive communication procedures, which entail the registration of a symphony of information in the subjective experience of feelings and desires. The essential function of this entails self-regulation and behavioral flexibility a.k.a. free will via the voices of the most basic units underpinning behavior and evolution: the genes, which are generally whispering in pianissimo, sometimes roaring in crescendo, but always insistent and present.

REFERENCES

1. Abelson, J. L., & Nesse, R. M., (1990). Cholecystokinin-4 and panic. *Archives of General Psychiatry, 47*(4), 395.
2. Amanzio, M., & Benedetti, F., (1999). Neuropharmacological dissection of placebo analgesia: Expectation-activated opioid systems versus conditioning-activated specific subsystems. *Journal of Neuroscience, 19*(1), 484–494.
3. Arletti, R., Benelli, A., & Bertolini, A., (1992). Oxytocin involvement in male and female sexual behavior. In: Pedersen, C. A., Caldwell, J. D., Jirikowski, G. F., & Insel, T. R., (eds.), *Oxytocin in Maternal, Sexual, and Social Behaviors* (Vol. 1, pp. 180–193). New York: New York Academy of Sciences.
4. Asberg, M., (1994). Monoamine transmitters in human aggressiveness and violence: A selected review. *Criminal Behavior and Mental Health, 4*(4), 303–327.
5. Bale, T. L., & Dorsa, D. M., (1995). Sex-differences in and effects of estrogen on oxytocin receptor messenger-ribonucleic-acid expression in the ventromedial hypothalamus. *Endocrinology, 136*(1), 27–32.
6. Basbaum, A. I., & Fields, H. L., (1978). Endogenous pain control mechanisms: Review and hypothesis. *Annals of Neurology, 4*(5), 451–462.
7. Basbaum, A. I., Clanton, C. H., & Fields, H. L., (1976). Opiate and stimulus produced analgesia: Functional anatomy of a medullospinal pathway. *Proceedings of the National Academy of Sciences of the United States of America, 73*(12), 4685–4688.
8. Basbaum, A. I., Fields, H. L., & Mandel, M. A., (1979). Endogenous pain control mechanisms: Review and hypothesis. *Plastic and Reconstructive Surgery, 64*(3), 431.
9. Beatty, J., (1995). *Principles of Behavioral Neuroscience* (Vol. 1, pp. 1–20). Madison, WI: Brown & Benchmark.
10. Belzung, C., Pineau, N., Beuzen, A., & Misslin, R., (1994). PD135158, a CCK-B antagonist, reduces "state," but not "trait" anxiety in mice. *Pharmacology, Biochemistry, and Behavior, 49*(2), 433–436.
11. Benkelfat, C., Bradwejn, J., Meyer, E., Ellenbogen, M., Milot, S., Gjedde, A., & Evans, A., (1995). Functional neuroanatomy of CCK4-induced anxiety in normal healthy volunteers. *American Journal of Psychiatry, 152(*8), 1180–1184.
12. Bolles, R. C., & Fanselow, M. S., (1982). Endorphins and behavior. *Annual Review of Psychology, 33*(1), 87–101.

13. Bongioanni, P., (1991). Platelet MAO activity and personality: An overview. *New Trends in Experimental and Clinical Psychiatry, 7(*1), 17–28.
14. Bortolato, M., & Shih, J. C., (2011). Behavioral outcomes of monoamine oxidase deficiency: Preclinical and clinical evidence. *International Review of Neurobiology, 100*, 13–42.
15. Bowers, M. B., Choi, D. C., & Ressler, K. J., (2012). Neuropeptide regulation of fear and anxiety: Implications of cholecystokinin, endogenous opiates, and neuropeptide Y. *Physiology and Behavior, 107*(5), 699–710.
16. Bradwejn, J., (1993). Neurobiological investigations into the role of cholecystokinin in panic disorder. *Journal of Psychiatry and Neuroscience, 18*(4), 178–188.
17. Bradwejn, J., Koszycki, D., & Shriqui, C., (1991). Enhanced sensitivity to cholecystokinin tetrapeptide in panic disorder: Clinical and behavioral findings. *Archives of General Psychology, 48*(7), 603–610.
18. Bradwejn, J., Koszycki, D., Annable, L., Couetoux Du, T. A., Van, M. H., Den, B. J., & Westenberg, H., (1994). The panicogenic effects of cholecystokinin-tetrapeptide are antagonized by L-365,260, a central cholecystokinin receptor antagonist, in patients with panic disorder. *Archives of General Psychiatry, 51*(6), 486–493.
19. Bradwejn, J., Koszycki, D., Paradis, M., Reece, P., Hinton, J., & Sedman, A., (1995). Effect of CI-988 on cholecystokinin tetrapeptide-induced panic symptoms in healthy volunteers. *Biological Psychiatry, 38*, 742–746.
20. Bradwejn, J., Zhou, Y., Koszycki, D., & Shlik, J., (2000). A double-blind, placebo-controlled study on the effects of Gotu kola (*Centella asiatica*) onacoustic startle response in healthy subjects. *Journal of Clinical Psychopharmacology, 20(*6), 680–684.
21. Brown, S. L., Botsis, A., & Van, P. H. M., (1994). Serotonin and aggression. *Journal of Offender Rehabilitation, 21*, 27–39.
22. Buck, R., & Van, L. C. A., (2002). Verbal and nonverbal communication: Distinguishing symbolic, spontaneous, and pseudo-spontaneous nonverbal behavior. *Journal of Communication, 52*(3), 522–541.
23. Buck, R., (1993a). What is this thing called subjective experience? Reflections on the neuropsychology of qualia. *Neuropsychology, 7*(4), 490–499.
24. Buck, R., (1993b). Spontaneous communication and the foundation of the interpersonal self. In: Neisser, U., (ed.), *The Perceived Self: Ecological and Interpersonal Sources of Self-Knowledge: Proceedings of the Emory Symposium on Cognition* (Vol. 1, pp. 216–36). Cambridge University

Press.

25. Caldwell, H. K., Dike, O. E., Stevenson, E. L., Storck, K., & Young, W. S. III., (2010). Social dominance in male vasopressin 1b receptor knockout mice. *Hormones and Behavior, 58*(2), 257–263.
26. Cardoso, C., Linnen, A., Joober, R., & Ellenbogen, M. A., (2012). Coping style moderates the effect of intranasal oxytocin on the mood response to interpersonal stress. *Experimental and Clinical Psychopharmacology, 20*(2), 84–91.
27. Carter, C. S., & DeVries, A. C., (1999). Stress and soothing: An endocrine perspective. In: Lewis, M., & Douglas, R., (eds.), *Soothing and Stress* (Vol. 1, pp. 3–18). Mahwah, NJ: Lawrence Erlbaum Associates.
28. Carter, C. S., Grippo, A. J., Pournajafi-Nazarloo, H., Ruscio, M. G., Porges, S. W., Inga, D. N., & Rainer, L., (2008). Oxytocin, vasopressin and sociality. *Progress in Brain Research, 170*, 331–336.
29. Castrogiovanni, P., Maremmani, I., Bongioanni, P., & Marazziti, D., (1990). Platelet monoaminooxidase activity and behavioral characteristics in humans. *Neuropsychobiology, 23*, 173–176.
30. Chhatwal, J. P., Gutman, A. R., Maguschak, K. A., Bowser, M. E., Yang, Y., Davis, M., & Ressler, K. J., (2009). Functional interactions between endocannabinoid and CCK neurotransmitter systems may be critical for extinction learning. *Neuropsychopharmacology, 34*(2), 509–521.
31. Coccaro, E. F., (1992). Impulsive aggression and central serotonergic system function in humans: An example of a dimensional brain-behavior relationship. *International Clinical Psychopharmacology, 7*(1), 3–12.
32. Costa, E., & Guidotti, A., (1991). Diazepam binding inhibitor (DBI): A peptide with multiple biological actions. *Life Sciences, 49*(5), 325–344.
33. Coursey, R. D., Buchsbaum, M. S., & Murphy, D. L., (1979). Platelet MAO activity and evoked potentials in the identification of subjects biologically at risk for psychiatric disorders. *British Journal of Psychiatry, 134,* 372–381.
34. Craig, I. W., (2005). The role of monoamine oxidase A, MAOA, in the aetiology of antisocial behavior: The importance of gene-environment interactions. In: Bock, G., & Goode, J., (eds.), *Molecular Mechanisms Influencing Aggressive Behaviors: The Novartis Foundation Symposium* (Vol. 268, pp. 227–237). Chichester: John Wiley.
35. De Beun, R., Geerts, N. E., Jansen, E., Slangen, J. L., & Van De, P. N. E., (1991). Luteinizing hormone-releasing hormone-induced conditioned place preference in male rats. *Pharmacology Biochemistry and Behavior, 39*(1), 143–147.

36. De Montigny, C., (1989). Cholecystokinin tetrapeptide induces panic-like attacks in healthy volunteers. *Archives of General Psychology, 46*(6), 511–517.

37. DeWied, D., & Gispen, W. H., (1977). Behavioral effects of peptides. In: Gainer, H., (ed.), *Peptides in Neurobiology* (Vol. 1, pp. 397–448). New York: Plenum.

38. Dixon, D., Jenkens, I., Moody, R. T. J., & Zhuravlev, A. Y., (2001). *Atlas of Life on Earth* (Vol. 1, pp. 1–23). New York: Barnes & Noble.

39. Dixon, T., (2012). "Emotion": The history of a keyword in crisis. *Emotion Review, 4*, 338–344.

40. Donaldson, Z. R., & Young, L. J., (2008). Oxytocin, vasopressin, and the neurogenetics of sociality. *Science, 322*(5903), 900–904.

41. Eccles, J., (1965). The synapse: From electrical to chemical transmission. *Scientific American, 212*, 56–66.

42. Farook, J. M., Zhu, Y. Z., Wang, H., Moochhala, S., Lee, L., & Wong, P. T., (2001). Strain differences in freezing behavior of PVG hooded and Sprague Dawley rats: Differential cortical expression of cholecystokinin-sub-2 receptors. *Neuroreport: For Rapid Communication of Neuroscience Research, 12(*12), 2717–2720.

43. Ferrarese, C., Appollonio, I., Bianchi, G., Frigo, M., Marzorati, C., Pecora, N., Perego, M., Pierpaoli, C., & Frattola, L., (1993). Benzodiazepine receptors and diazepam binding inhibitor: A possible link between stress, anxiety, and the immune system. *Psychoneuroendocrinology, 18*(1), 3–22.

44. Ferris, C. F., Stolberg, T., Kulkarni, P., Murugavel, M., Blanchard, R., Blanchard, D. C., Febo, M., et al., (2008). Imaging the neural circuitry and chemical control of aggressive motivation. *BMC Neuroscience, 9*, 111.

45. Fischman, M. W., (1984). The behavioral pharmacology of cocaine in humans. In: Grabowski, J., (ed.), *Cocaine: Pharmacology, Effects and Treatment of Abuse* (Vol. 1, pp. 73–92). Washington, DC: US Government Printing Office.

46. George, D. T., Umhau, J. C., Phillips, M. J., Emmela, D., Ragan, P. W., Shoaf, S. E., & Rawlings, R. R., (2001). Serotonin, testosterone and alcohol in the etiology of domestic violence. *Psychiatry Research, 104*(1), 27–37.

47. Gingrich, B. S., Huot, R. L., Wang, Z., & Insel, T. R., (1997). Differential Fos expression following microinjection of oxytocin or vasopressin in the prairie vole brain. In: Carter, C. S., Lederhendler, I. I., & Kirkpatrick, B., (eds.), *The Integrative Neurobiology of Affiliation, Annals of the New York Academy of Sciences* (Vol. 807, pp. 504, 505). New York: New York

Academy of Sciences.

48. Gore, R., (2003). The rise of mammals. *National Geographic, 203*(4), 2–37.
49. Grunberg, N. E., & Straub, R. O., (1992). The role of gender and taste class in the effects of stress on eating. *Health Psychology, 11*, 97–100.
50. Guarneri, P., Guidotti, A., & Costa, E., (1990). On the processing of diazepam binding inhibitor (DBI) in human brain. In: Biggio, G., & Costa, E., (eds.), *Advances in Biochemical Pharmacology: GABA and Benzodiazepine Receptor Subtypes: Molecular Biology, Pharmacology, and Clinical Aspects* (Vol. 46, pp. 201–111). New York: Raven Press.
51. Guastella, A. J., & MacLeod, C., (2012). A critical review of the influence of oxytocin nasal spray on social cognition in humans: Evidence and future directions. *Hormones and Behavior, 61*(3), 410–418.
52. Guastella, A. J., Mitchell, P. B., & Mathews, F., (2008). Oxytocin enhances the encoding of positive social memories in humans. *Biological Psychiatry, 64*(3), 256–558.
53. Guerrero, L. K., Trost, M. R., & Yoshimura, S. M., (2005). Romantic jealousy: Emotions and communicative responses. *Personal Relationships, 12*(2), 233–252.
54. Gur, R. C., Mozley, L. H., Mozley, P. D., Resnick, S. M., Karp, J. S., Alavi, A., Arnold, S. E., & Gur, R. E., (1995). Sex differences in regional cerebral glucose metabolism during a resting state. *Science, 267*(5197), 528–531.
55. Haenisch, B., Bilkei-Gorzo, A., Caron, M. G., & Bönisch, H., (2009). Knockout of the norepinephrine transporter and pharmacologically diverse antidepressants prevent behavioral and brain neurotrophin alterations in two chronic stress models of depression. *Journal of Neurochemistry, 111*(2), 403–416.
56. Hahn, T., Dresler, T., Ehlis, A., Plichta, M. M., Heinzel, S., Polak, T., Lesch, K. P., et al., (2009). Neural response to reward anticipation is modulated by gray's impulsivity. *Neuroimage, 46*(4), 1148–1153.
57. Haidt, J., & Keltner, D., (1999). Culture and facial expression: Open-ended methods find more faces and a gradient of recognition. *Cognition & Emotion, 13*, 225–226.
58. Haidt, J., (2001). The emotional dog and its rational tail: A social intuitionist approach to moral judgment. *Psychological Review, 108*(4), 814–834.
59. Hall, E. T., (1966). *The Hidden Dimension* (Vol. 6, No. 1, p. 94). Garden City, NY: Doubleday (1973); The Hidden Dimension, Leonardo.

60. Hebb, A. L., Zacharko, R. M., Dominguez, H., Trudel, F., Laforest, S., & Drolet, G., (2002). Odor-induced variation on anxiety-like behavior in mice is associated with discrete and differential effects on mesocorticolimbic cholecystokinin mRNA expression. *Neuropsychopharmacology, 27*(5), 744–755.
61. Hebb, D. O., (1949). *The Organization of Behavior* (Vol. 1, pp. 1–20). New York: John Wiley.
62. Hertenstein, M. J., Verkamp, J. M., Kerestes, A. M., & Holmes, R. M., (2006). The communicative functions of touch in humans, nonhuman primates, and rats: A review and synthesis of the empirical research. *Genetic, Social, and General Psychology Monographs, 132*(1), 5–94.
63. Hess, W. R., (1928). Stammganglien-reizversuche [Brainstem stimulation attempts]. *Berliner Gesellschaft fur Physiologie, 42,* 554.
64. Hughes, J., Smith, T., Morgan, B., & Fothergill, L., (1975). Purification and properties of enkephalin: The possible endogenous ligand for the morphine receptor. *Life Sciences, 16*(12), 1753–1758.
65. Insel, T. R., (1992). Oxytocin: A neuropeptide for affiliation: Evidence from behavioral, receptor autoradiographic, and comparative studies. *Psychoneuroendocrinology, 17*(1), 3–35.
66. Insel, T. R. (2002). Implications for the neurobiology of love. *Altruism & altruistic love: Science, philosophy, and religion in dialogue*, 254-263.
67. Iversen, L., (1979). The chemistry of the brain. *Scientific American, 241*(3), 134–149.
68. Jaffe, J. H., (1985). Drug addiction and drug abuse. In: Gilman, A. G., Goodman, L. S., Rail, T. W., & Murad, F., (eds.), *Goodman and Gilman's The Pharmacologic Basis of Therapeutics* (Vol. 7, pp. 532–581). New York: Macmillan.
69. Janssen, S. A., & Arntz, A., (2001). Real-life stress and opioid-mediated analgesia in novice parachute jumpers. *Journal of Psychophysiology, 15*(2), 106–113.
70. Javanmard, M., Shlik, J., Kennedy, S. H., Vaccarino, F. J., Houle, S., & Bradwejn, J., (1999). Neuroanatomic correlates of CCK-4-induced panic attacks in healthy humans: A comparison of two time points. *Biological Psychiatry, 45*(7), 872–882.
71. Jerabek, I., Boulenger, J. P., Lavallee, Y. J., & Jolicoeur, F. B., (1998). Psychological predictors of CCK-sub-4-induced panic-like symptoms. *Human Psychopharmacology Clinical and Experimental, 13*(1), 35–42.
72. Kandell, S., Schwarts, J., & Jessell, T., (2000). *Principles of Neural*

Science (Vol. 4, p. 2–32), New York: McGraw-Hill.

73. Katsura, M., Mohri, Y., Shuto, K., Tsujimura, A., Ukai, M., & Ohkuma, S., (2002). Psychological stress, but not physical stress, causes increase in diazepam binding inhibitor (DBI) mRNA expression in mouse brains. *Molecular Brain Research, 104*(1), 103–109.

74. Kavaliers, M., & Hirst, M., (1986). An octadecaneuropeptide (ODN) derived from diazepam binding inhibitor increases aggressive interactions in mice. *Brain Research, 383*(1, 2), 343–349.

75. Kennedy, J. L., Bradwejn, J., Koszycki, D., King, N., Crowe, R., Vincent, J., & Fourie, O., (1999). Investigation of cholecystokinin system genes in panic disorder. *Molecular Psychiatry, 4*(3), 284, 285.

76. Kohler, E., Keysers, C., Umiltà, M. A., Fogassi, L., Gallese, V., & Rizzolatti, G., (2002). Hearing sounds, understanding actions: Action representation in mirror neurons. *Science, 297*(5582), 846–848.

77. Kolak, A. M., & Volling, B. L., (2011). Sibling jealousy in early childhood: Longitudinal links to sibling relationship quality. *Infant and Child Development, 20*(2), 213–226.

78. Kolber, B. J., Boyle, M. P., Wieczorek, L., Kelley, C. L., Onwuzurike, C. C., Nettles, S. A., Vogt, S. K., & Muglia, L. J., (2010). Transient early-life forebrain corticotropin-releasing hormone elevation causes long-lasting anxiogenic and despair-like changes in mice. *Journal of Neuroscience, 30(7*), 2571–2581.

79. Kolmer, M., Roos, C., Tirronen, M., Myöhänen, S., & Alto, H., (1994). Tissue-specific expressions of the diazepam-binding inhibitor in drosophila melanogaster: Cloning, structure, and localization of the genes. *Molecular Cell Biology, 14*(10), 6983–6995.

80. Koolhaas, J. M., Van, D. B. T. H. C., Roozendaal, B., & Boorsma, F., (1990). Medial amygdala and aggressive behavior: Interaction between testosterone and vasopressin. *Aggressive Behavior, 16*(3), 223–229.

81. Koscik, T. R., & Tranel, D., (2011). The human amygdala is necessary for developing and expressing normal interpersonal trust. *Neuropsychologia, 49*(4), 602–611.

82. Koszycki, D., Cox, B. J., & Bradwejn, J., (1993). Anxiety sensitivity and response to cholecystokinin tetrapeptide in healthy volunteers. *American Journal of Psychiatry, 150*(12), 1881–1883.

83. Lam, D., & Marteleto, L., (2005). Small families and large cohorts: The impact of the demographic transition on schooling in Brazil. In: Lloyd, C., Behrman, J., Stromquist, N., & Cohen, B., (eds.), *The Changing Transitions to Adulthood in Developing Countries: Selected Studies* (Vol.

1, pp. 56–83). Washington, DC: National Academy of Sciences.

84. Landis, C., (1924). Studies of emotional reactions: General behavior and facial expression. *Journal of Comparative Psychology, 4*(5), 447–510.
85. Lane, R. D., Chua, P. M., & Dolan, R. J., (1999). Common effects of emotional valence, arousal and attention on neural activation during visual processing of pictures. *Neuropsychologia, 37*(9), 989–997.
86. Langer, E., Blank, A., & Chanowitz, B., (1978). The mindlessness of ostensibly thoughtful action: The role of placebic information in interpersonal interaction. *Journal of Personality and Social Psychology, 36*(6), 635–642.
87. LeRoith, D., Liotta, A. S., Roth, J., Shiloach, J., Lewis, M. E., Pert, C. B., & Krieger, D. T., (1982). Corticotropin and β-endorphin-like materials are native to unicellular organisms. *Proceedings of the National Academy of Sciences of the United States of America, 79*(6), 2086–2090.
88. LeRoith, D., Shiloach, J., Roth, J., & Lesniak, M. A., (1980). Evolutionary origins of vertebrate hormones: Substances similar to mammalian insulin are native to unicellular eukaryotes. *Proceedings of the National Academy of Sciences of the United States of America, 77*(10), 6184–6188.
89. Levine, J. D., Gordon, N. C., Jones, R. T., & Fields, H. L., (1978). The narcotic antagonist naloxone enhances clinical pain. *Nature, 272*, 826–827.
90. Lewejohann, L., Kloke, V., Heiming, R. S., Jansen, F., Kaiser, S., Schmitt, A., Lesch, K. P., & Sachser, N., (2010). Social status and day-to-day behavior of male serotonin transporter knockout mice. *Behavioral Brain Research, 211*(2), 220–228.
91. Lindquist, K. A., Wager, T. D., Kober, H., Bliss-Moreau, E., & FeldmanBarrett, L., (2012). The brain basis of emotion: A meta-analytic review. *Behavioral and Brain Sciences, 35*(3), 121–143.
92. Lindsley, D. B., (1951). Emotion. In: Stevens, S. S., (ed.), *Handbook of Experimental Psychology* (Vol. 1, pp. 473–516). New York: John Wiley.
93. Loumaye, E., Thorner, J., & Catt, K. J., (1982). Yeast mating pheromone activates mammalian gonadotropins: Evolutionary conservation of a reproductive hormone? *Science, 218*(4579), 1323–1325.
94. Macbeth, A. H., Stepp, J. E., Lee, H. J., Young, W. S. III., & Caldwell, H. K., (2010). Normal maternal behavior, but increased pup mortality, in conditional oxytocin receptor knockout females. *Behavioral Neuroscience, 124*(5), 677–85.
95. Margules, L., & Sagan, D., (1995). *What is Life?* (Vol. 1, pp. 1–22). New

York: Simon & Schuster.

96. Martínez, I., Rosa, M., Arsuaga, J. L., Jarabo, P., Quam, R., Lorenzo, C., Gracia, A., et al., (2004). Auditory capacities in Middle Pleistocene humans from the Sierra de Atapuerca in Spain. *Proceedings of the National Academy of Sciences of the United States of America, 101*(27), 9976–9981.
97. Marx, J. L., (1985). "Anxiety peptide" found in brain. *Science, 227*(4689), 934.
98. Matthews, G. G., (1998). *Neurobiology: Molecules, Cells, and Systems* (Vol. 2, pp. 2–12). Oxford: Blackwell Science.
99. Mayer, D. J., Price, D. D., & Rafii, A., (1977). Antagonism of acupuncture analgesia in man by the narcotic antagonist naloxone. *Brain Research, 121*(2), 368–372.
100. McCarthy, M. M., (1990). Oxytocin inhibits infanticide in female house mice (Mus domesticus). *Hormones and Behavior, 24*(3), 365–375.
101. Melzack, R., & Wall, P. D., (1965). Pain mechanisms: A new theory. *Science, 150*(3699), 971–979.
102. Meredith, M., & Fernandez-Fewell, G., (1994). Vomeronasal system, LHRH, and sex behavior. *Psychoneuroendocrinology, 19*(5–7), 657–672.
103. Miller, W. L., Baxter, J. D., & Eberhardt, N. L., (1983). Peptide hormone genes: Structure and evolution. In: Krieger, D. T., & Martin, J. B., (eds.), *Brain Peptides* (Vol. 1, pp. 15–78). New York: John Wiley.
104. Moskowitz, D. S., Pinard, G., Zuroff, D. C., Annable, L., & Young, S. N., (2001). The effect of tryptophan on social interaction in everyday life: A placebo-controlled study. *Neuropsychopharmacology, 25*, 277–289.
105. Murphy, D. L., (1973). Technical strategies for the study of catecholamines in man. In: Usdin, E., & Snyder, S., (eds.), *Frontiers in Catecholamine Research* (pp. 1077–1082). Oxford: Pergamon.
106. New, A. S., Hazlett, E. A., Buchsbaum, M. S., Goodman, M., Reynolds, D., Mitropoulou, V., Sprung, L., et al., (2002). Blunted prefrontal cortical[18] fluorodeoxyglucose positron emission tomography response to meta-chlorophenyl piperazine in impulsive aggression. *Archives of General Psychiatry, 59*(7), 621–29.
107. Niall, H. D., (1982). The evolution of peptide hormones. *Annual Review of Physiology, 44*(1), 615–624.
108. Owens, G. P., Sinha, A. K., Sikela, J. M., & Hahn, W. E., (1989). Sequence and expression of the murine diazepam binding inhibitor. *Molecular Brain Research, 6*, 101–108.

109. Panksepp, J., (1981). Hypothalamic integration of behavior. In Morgane, P., & Panksepp, J., (eds.), *Handbook of the Hypothalamus: Behavioral Studies of the Hypothalamus* (Vols. 1, 3, pp. 289–437). New York: Marcel Bekker.

110. Panksepp, J., (1993). Neurochemical control of moods and emotions: Amino acids to neuropeptides. In: Lewis, M., & Haviland, J., (eds.), *Handbook of Emotions* (pp. 87–107). New York: Guilford Press.

111. Panksepp, J., (1996). The roles of brain opioids, oxytocin and prolactin in social affect and social bonding. *Presentation at the New York Academy of Sciences Conference on the Integrative Neurobiology of Affiliation.* Georgetown University, Washington, DC.

112. Panksepp, J., (1998). *Affective Neuroscience: The Foundations of Human and Animal Emotions* (Vol. 1, pp. 4–22). New York: Oxford University Press.

113. Pederson, C. A., (1997). Oxytocin control of maternal behavior: Regulation by sex steroids and offspring stimuli. In: Carter, C. S., Lederhendler, I. I., & Kirkpatrick, B., (eds.), *The Integrative Neurobiology of Affiliation, Annals of the New York Academy of Sciences* (Vol. 807, pp. 126–45). New York: New York Academy of Sciences.

114. Pert, C. B., & Snyder, S. H., (1973). Opiate receptor: Demonstration in nervous tissue. *Science, 179*(4077), 1011–1014.

115. Pert, C., (1985). *Neuropeptides and Their Receptors: Substrates for the Biochemistry of Emotion*. Keynote address at the inaugural meeting of the International Society for Research on Emotions. Harvard University, Cambridge, MA.

116. Pert, C., (1997). *Molecules of Emotion: Why You Feel the Way You Feel.* New York. Scribner (Vol. 1, pp. 1–23).

117. Pinel, J. P. J., (1993). *Biopsychology* (2nd edn., Vol. 1, pp. 4–21). Needham Heights, MA: Allyn & Bacon.

118. Pitkow, L. J., Sharer, C. A., Ren, X., Insel, T. R., Terwilliger, E. F., & Young, L. J., (2001). Facilitation of affiliation and pair-bond formation by vasopressin receptor gene transfer into the ventral forebrain of a monogamous vole. *Journal of Neuroscience, 21*(18), 7392–7396.

119. Potegal, M., & Ferris, C. F., (1989). Intraspecific aggression in male hamsters is inhibited by intrahypothalamic vasopressin-receptor antagonist. *Aggressive Behavior, 15*(4), 311–320.

120. Purves, D., Augenstine, G. J., Fitzpatrick, D., Katz, L. C., LaMantia, A. S., & McNamara, J. O., (1997). *Neuroscience* (Vol. 1, pp. 10–20). Sunderland, MA: Sinauer.

121. Rampello, L., Alvano, A., Chiechio, S., Raffaele, R., Vecchio, I., & Malaguarnera, M., (2005). An evaluation of efficacy and safety of reboxetine in elderly patients affected by "retarded" post-stroke depression: A random, placebo-controlled study. *Archives of Gerontology and Geriatrics, 40*(3), 275–285.

122. Redmond, Jr. D. E., & Murphy, D. L., (1975). Behavioral correlates of platelet monoamine oxidase (MAO) activity in rhesus monkeys. *Psychosomatic Medicine, 37*, 80.

123. Redmond, Jr. D. E., Murphy, D. L., & Baulu, J., (1979). Platelet monoamine oxidase activity correlates with social affiliative and agonistic behaviors in normal rhesus monkeys. *Psychosomatic Medicine, 41*, 87–100.

124. Reiman, E. M., Fusselman, M. J., Fox, P. T., & Raichle, M. E., (1989). Neuroanatomical correlates of anticipatory anxiety. *Science, 243*(4894), 1071–1074.

125. Reti, I. M., Xu, J. Z., Yanofski, J., McKibben, J., Uhart, M., Cheng, Y. J., Zandi, P., & Nestadt, G., (2011). Monoamine oxidase A regulates antisocial personality in whites with no history of physical abuse. *Comprehensive Psychiatry, 52*(2), 188–194.

126. Richard, P., Moos, F., & Freund-Mercier, M. J., (1991). Central effects of oxytocin. *Physiological Reviews, 71*(2), 331–370.

127. Risch, S. C., (1991). *Central Nervous System Peptide Mechanisms in Stress and Depression* (Vol. 1, pp. 1–22). Washington, DC: American Psychiatric Press.

128. Rockliff, H., Karl, A., McEwan, K., Gilbert, J., Matos, M., & Gilbert, P., (2011). Effect of oxytocin on compassion-focused imagery. *Emotion, 11*, 1388–1396.

129. Rodriguez De, F. F., Carrera, M., Rocio, A., Navarro, M., Koob, G. F., & Weiss, F., (1997). Activation of corticotropin-release factor in the limbic system during cannabinoid withdrawal. *Science, 276*(5321), 2050–2054.

130. Ross, E. D., Homan, R. W., & Buck, R., (1994). Differential hemispheric lateralization of primary and social emotions. *Neuropsychiatry, Neuropsychology, and Behavioral Neurology, 7*(1), 1–19.

131. Roth, B. L., (1994). Multiple serotonin receptors: Clinical and experimental aspects. *Annals of Clinical Psychiatry, 6*, 67–78.

132. Schally, A. V., Coy, D. H., Arimura, A., Redding, T. W., Kastin, A. J., Meyers, C., & Millar, R., (1978). Hypothalamic peptide hormones and their analogues. In: Cox, B., Morris, I. D., & Weston, A. H., (eds.), *Pharmacology of the Hypothalamus* (pp. 161–201). Baltimore: University Park Press.

133. Sher, K. J., Bylund, D. B., Walitzer, K. S., Hartmann, J., & Ray-Prenger, C., (1994). Platelet monoamine oxidase (MAO) activity: Personality, substance use, and the stress-response-dampening effect of alcohol. *Experimental and Clinical Psychopharmacology, 2*(1), 53–81.

134. Siever, L., & Trestman, R. L., (1993). The serotonin system and aggressive personality disorder. *International Clinical Psychopharmacology, 8*(2), 33–39.

135. Ter, R. G., De Craen, A. J. M., De Boer, A., & Kessels, A. G. H., (1998). Is placebo analgesia mediated by endogenous opioids? *Pain: A Systematic Review, 76*(3), 273–275.

136. Thompson, R. R., George, K., Walton, J. C., Orr, S. P., & Benson, J., (2006). Sex-specific influences of vasopressin on human social communication. *Proceedings of the National Academy of Sciences of the United States of America, 103*(20), 7889–94.

137. Tse, W. S., & Bond, A. J., (2002a). Difference in serotonergic and noradrenergic regulation of human social behaviors. *Psychopharmacology, 159*(2), 216–221.

138. Tse, W. S., & Bond, A. J., (2002b). Serotonergic intervention affects both social dominance and affiliative behavior. Psychopharmacology, 161(3), 324–330.

139. Tsutsumi, T., Akiyoshi, J., Hikichi, T., Kiyota, A., Kohno, Y., Katsuragi, S., & Nagayama, H., (2001). Suppression of conditioned fear by administration of CCKB receptor antisense oligodeoxynucleotide into the lateral ventricle. *Pharmacopsychiatry, 34*(6), 232–237.

140. Tuinier, S., Verhoeven, W. M., & Van, P. H. M., (1995). Cerebrospinal fluid 5-hydroxyindolacetic acid and aggression: A critical reappraisal of the clinical data. *International Clinical Psychopharmacology, 10(*3), 147–156.

141. Post, S. G., Underwood, L. G., Schloss, J. P., & Hurlbut, W. B. (2002). *Altruism and altruistic love: Science, philosophy, and religion in dialogue*. Oxford University Press (Vol. 1, pp. 254–263). Oxford University Press.

142. Uzefovsky, F., Shalev, I., Israel, S., Knafo, A., & Ebstein, R. P., (2012). Vasopressin selectively impairs emotion recognition in men. *Psychoneuroendocrinology, 37*(4), 576–580.

143. Walker, R. J., Brooks, H. L., & Holden-Dye, L., (1996). Evolution and overview of classical transmitter molecules and their receptors. *Parasitology, 113*, 3–33.

144. Walker, W. F., (1987). *Functional Anatomy of the Vertebrates: An*

Evolutionary Perspective (Vol. 1, pp. 1–22). Philadelphia: Saunders.

145. Windslow, J. T., Hastings, N., Carter, C. S., Harbaugh, C. R., & Insel, T. R., (1993). A role for central vasopressin in pair bonding in the monogamous prairie vole. *Nature, 365*(6446), 545–548.
146. Wise, R. A., & Bozarth, M. A., (1987). A psychomotor stimulant theory of addiction. *Psychological Review, 94*(4), 469–492.
147. Yanase, H., Shimizu, H., Yamada, K., & Iwanaga, T., (2002). Cellular location of the diazepam binding inhibitor in glial cells with special reference to its coexistence with brain-type fatty acid-binding protein. *Archives of Histology and Cytology, 65*(1), 27–36.
148. Zuckerman, M., (1993). P-impulsive sensation seeking and its behavioral, psychophysiological, biochemical correlates. *Neuropsychobiology, 28*(1, 2), 30–36.
149. Zuckerman, M., Buchsbaum, M. S., & Murphy, D. L., (1980). Sensation seeking and its biological correlates. *Psychological Bulletin, 88*(1), 187–214.

CHAPTER 3 Architecture and Neurochemistry of the Central Complex

Contents

3.1. INTRODUCTION

Known as the central complex, it is one of the most prominent neuroarchitectures to be discovered in the pest brain, and it is responsible for a wide range of activities like as propulsion, stridulatory activity, spatial orientation, and spatial memory. Using the grasshopper as a model insect system, we demonstrate that the complicated wiring of the fan-shaped body (FB) begins early in embryogenesis when the first progeny of four protocerebral stem cells (called W, X, Y, and Z) in each brain hemisphere establish a set of tracts leading to the primary commissural system (also known as the primary commissural system). Once this occurs, the FB begins to develop a columnar neuroarchitecture, which is completed throughout embryogenesis, by decussating subset-specific commissural neurons at stereotypical places within the midline of the brain. According to the results of an investigation on the expression patterns of several neurochemicals in the central complex, including neuropeptides, a neurotransmitter (NT), and the gas nitric oxide (NO), they appear gradually and, in a substance, -specific manner throughout embryogenesis (Ito et al., 1997; Meinertzhagen et al., 1998). Each neuroactive substance is produced by neurons situated at stereotypical places in a specific central complex lineage, demonstrating that the stem cells are biochemically multipotent in their ability to differentiate. The structure of axons expressing the various neurochemicals inside the central complex is topologically connected to the placement of the neurons within the lineages, and hence to their birthdates, within the central complex. There are several layers to the neurochemical expression designs in the FB, which corresponds to the temporal architecture found in the lineages. These findings may give insight into the evolution of adaptive behaviors since they establish a link between neural anatomy and neurochemical organization in central complex (Strausfeld, 1999; Farris and Sinakevitch, 2003; Ito and Awasaki, 2008).

Insect MB (mushroom bodies) and central complexes offer a perfect environment for research aimed at understanding the evolving/genetic foundation of neural function and behavior (Wegerhoff and Breidbach, 1992; Mizunami et al., 1997). The pest central complex, which is found in the midline of the brain, is made up of five main components: the fan-shaped body (PB), protocerebral bridge, ellipsoid body, lateral accessory lobes (LAL), and noduli (N), and is one of the most unique neuroarchitectures among arthropods (Neuser et al., 2008; Pan et al., 2009; Harley and Ritzmann, 2010).

Both mutation analysis and lesioning investigations have established the functions that these distinct modules show in coordinating motor orientation, activity, and visual pattern storage and memory (Huber, 1960; Ilius et al., 1994; Strauss, 2002; Liu et al., 2006).

The FB is the most visible central complex module in insects, with layers of stereotypic columnar structure and dendritic arbors, both created by axonal predictions of masses of neurons in the PI (pars intracerebral) area of every protocerebrally hemisphere. It has been suggested that the degree to which an insect's central complex is developed is connected to the species' activity (Strausfeld, 2012). According to developmental research, the complicated columnlike wiring of the FB is created quickly, although at different stages depending on the insect's lifestyle. It builds fully through embryogenesis in the grasshopper (Boyan et al., 2008a, b), however modules are improved sequentially through larval development in beetles, and early-born neurons developed interconnected into the central complex during the pupal to adult transition in flies (Renn et al., 1999). Divergent to this temporal variety, the growing mechanisms at work are thought to be preserved, and they involve a process known as fascicle switching, in which the afferents of commissural neurons systematically intersectant at stereotypic locations throughout the midline of the brain, resulting in columnar neuroarchitecture that is required for adaptive behavior (Reichert and Boyan, 2011).

The evolution of the neurochemical design of the central complex occurs in conjunction with the development of the central complex's anatomical structure. This is less well known, but it is crucial for understanding the function that core complicated circuits play in human behavior when studying the brain. Although neuroactive chemicals are expressed by grasshopper neural stem cells while they are still extant, this allows the contributing neurons to be ontogenetically recognized as per the lineage of origin from whence they originated (Boyan and Liu, 2014). As a result of this characteristic, central complex neuroblasts have been demonstrated to be multipotent, in that they may form lineages that contain a diverse array of neuroactive chemicals (Boyan et al., 2010a, b; Kahsai and Winther, 2011). The expression of such substances in the adult brain of Drosophila, on the other hand, has only been evidenced in cell clusters of the adult brain (Ito and Awasaki, 2008).

The grasshopper is used as a typical system for central generating new knowledge in this chapter. The FB's neuroarchitecture is linked to stereotypic axogenesis patterns including subsets of neurons from recognized neuroblasts. We then display those neuroactive chemicals required for synaptic connections in this system emerge gradually through development, following a sequential topology that links the location of neurons inside one lineage to their axonal and age project design. Temporal topology connects the structural and biochemical neuroarchitecture of the FB, potentially revealing information about the evolution of adaptive behavior (Williams, 1975; Heinze and Homberg, 2008).

3.2. NEURONAL CLASSES AND NEUROARCHITECTURE BEHAVIOR

The central complex is made up of five neuropilar components: PB, noduli (N), EB, FB, and LAL (lower division of the central body in the grasshopper), with the unpaired FB having the most prominent columns and stratified layers (Figure 3.1; Strausfeld, 2012; Pfeiffer and Homberg, 2014). Depending upon their projection patterns into and out of the grasshopper's numerous modules, four primary neuronal types have been identified in the central complex. These are, in a nutshell: (i) Columnar neurons are neurons that link single PB and/or FB columns to the LAL or noduli. The four fiber routes known as the w, x, y, and z tracts project columnar axons from the PB to the FB (Figure 3.1(B); Strausfeld, 1976; Hanesch et al., 1989; Williams and Boyan, 2008). These axons form nine columnar bundles inside the FB, which will be discussed in this chapter in terms of developmental expression of several neurochemicals. (ii) Tangential neurons (Homberg, 1991; Müller et al., 1997), which innervate single layers of the PB, EB, or FB and connect the central complex to other brain areas. (iii) Pontine neurons are inherent components within the FB that link certain columns and layers (Hanesch et al., 1989). (iv) Amacrine neurons, which are found only in the frontal lobe. The only known case contains a soma in the PI, a neurite in the z region, and arborizations in the FB's lateral hemisphere (Figure 3.2) (Armstrong and Young, 2010a, b; Riebli et al., 2013).

The insect central compound has been characterized as a multisensory neuropile that processes visual, olfactory signals, and mechanosensory on the one hand, even whereas portion as a premotor center that regulates acoustic communication, flight, walking, and wooing on the other, according to previous research. While certain sub-compartments of the EB are concerned in diverse characteristics of landmark and spatial learning, flight control, and orientation in Drosophila, the FB adjusts locomotory activity for negotiating barriers in the cockroach (Bender et al., 2010; Harley and Ritzmann, 2010). In the grasshopper brain, of all the core complex modules, the FB is likely the one that shows the most consistent relationship between neuronal morphology, overall neuroarchitecture, and information processing subserving behavior. This is accomplished using a specific type of visual information processing, termed sky polarization, which is essential to the overall navigation behavior of insects (see Wehner, 1989; Homberg et al., 2011; Weir and Dickinson, 2012). Electrical recordings from columnar (CPU1) neurons and identified tangential (TB1) in the PI area demonstrate preferential polarization sensitivities that are split into eight channels per brain hemisphere, as seen in the sky chart below (Figure 3.3(C)). Comparison of preferred polarization sensitivities in the PI with those in an upper layer of the FB indicates that there is a transformation en route that reflects the wiring plan for fibers entering the FB through the chiasmal system of w, x, and y tracts (see Figure 3.3(B) and Williams, 1975 for more information).

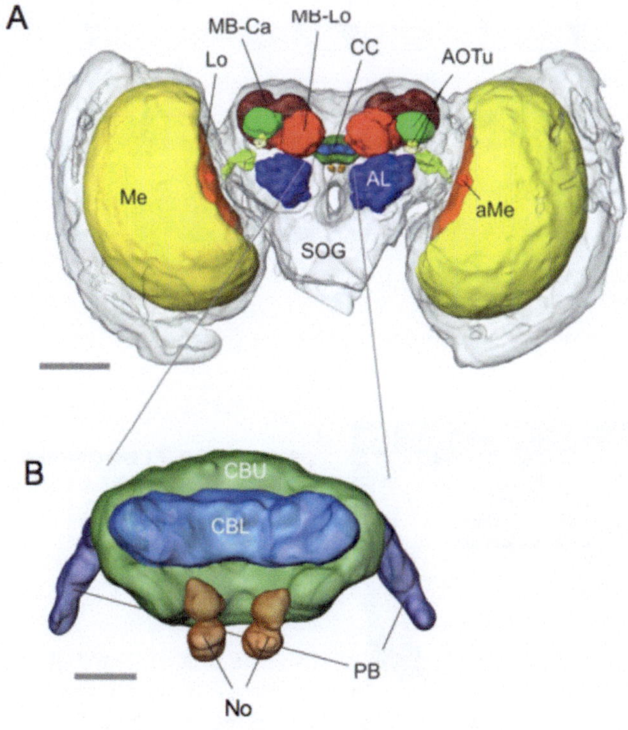

Figure 3.1. *Three-dimensional organization of the (a) adult monarch brain; and (b) central complex.*

Note: (A) A three-dimensional reconstruction of the main brain neuropils based on confocal image stacks (scanned with a 10x objective) generated from anti-synapsin stained wholemount preparations (voxel size: 3 m³). The surface of the brain is depicted in gray. (B) The identical reconstruction as in A, but from the front. Based on confocal image stacks (scanned with a 25x objective) acquired from anti-synapsin stained wholemount preparations (voxel size: 1 m³), a three-dimensional reconstruction of the key core complex components was created (Strausfeld, 2012).

Source: https://www.umassmed.edu/reppertlab/migration/the-central-complex/.

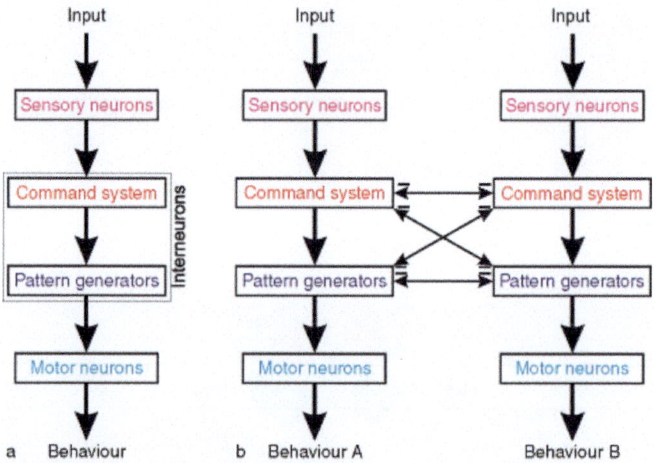

Figure 3.2. *Neural basics of behavior.*

Source: https://www.sciencedirect.com/topics/psychology/neural-basis-of-behavior.

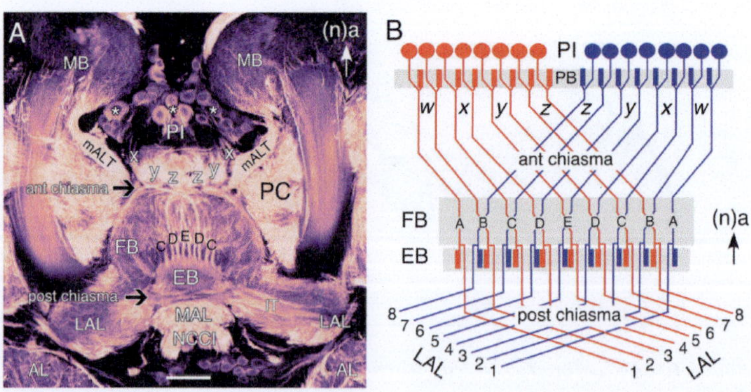

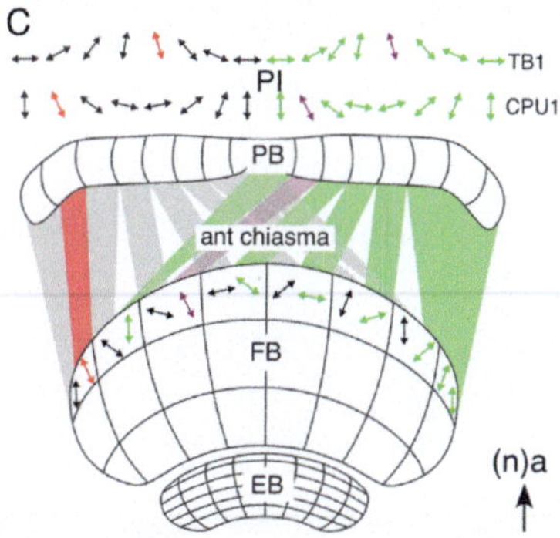

Figure 3.3. *Wiring of the central complex subserves information.*

Note: (A) The neuroarchitecture of the central complex seen among bilateral MB (mushroom bodies) in the central brain is revealed in a confocal picture of a brain slice in an adult grasshopper subsequent eight B7 immunolabeling. Axons from the pars intracerebral (PI) are directed into a chiasmal system (black arrow) anterior to the fan-shaped body through the z, x, y, and w areas (FB). These fibers subsequently form columnar bundles in the FB and project to the ellipsoid body (EB) and subsequently in the isthmus tract (IT) to the lateral accessory lobes through a posterior chiasmal system (black arrow) (LAL). The FB is also known as the upper section of the central body, while the EB is known as the lower division. White arrow indicates to frontal (a) as per neuraxis (n) and is repeated for emphasis in all panels. (B) Wiring diagram (not to scale) depicting the important plan of axon projections from PI neurons into the w, x, y, and z tract systems of the left (red) and right (blue) protocerebral hemispheres, and thus into the anterior chiasmal system to form nine columns (A–E) in the FB and EB of the central brain via the protocerebral bridge (PB). Axons exit the EB posteriorly and project to the bilateral LALs through a posterior chiasmal pathway. (C) A schematic depicts the preferred polarization sensitivities (double arrows) of columnar (CPU1) and tangential (TB1) neurons in the grasshopper's PI, as well as how these sensitivities are projected through the anterior chiasmal system to be signified inside the FB's columnar neuroarchitecture. mALT stands for medial antennal lobe tract; AL stands for antennal lobe; MAL stands for median accessory lobe; NCCI stands for nervus corporis cardiaci I; PC stands for protocerebrum. In (A), the scale bar equals 100 mm. Panelists (A) (Müller et al., 1997; Heinze and Homberg, 2008).

Source: https://www.researchgate.net/figure/Wiring-of-the-central-complex-subserves-information-processing-A-Confocal-image-of-a_fig10_307475381.

3.3. DEVELOPMENT OF THE CENTRAL COMPLEX

3.3.1. Organization of Neural Stem Cells

The central complex is a protocerebral neuromere that belongs to the brain's protocerebral neuromere (see Strausfeld, 2012). The brain of a grasshopper is made up of around 100 bilaterally symmetrical pairs of neural stem cells termed neuroblasts (Figure 3.4(A)), each of which is uniquely defined by molecular, positional, and temporal signals (Zacharias et al., 1993; Reichert and Boyan, 1997; Urbach and Technau, 2003).

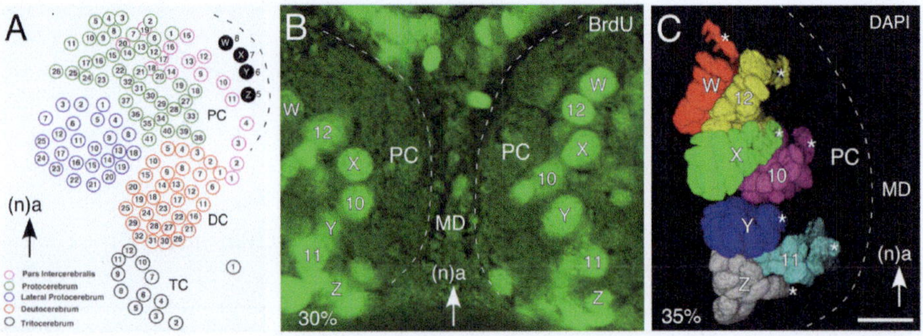

Figure 3.4. *In the grasshopper's early embryonic brain, neural stem cells are organized.*

Note: (A) Prior to mid-embryogenesis, a schematic illustrates the position of detected neuroblasts in the tritocerebrum (TC), deutocerebrum (DC), and protocerebrum (PC) of the grasshopper brain. W, X, Y, and Z neuroblasts produce offspring that contribute to the columnar arrangement of the central complex's FB. The arrow in all panels points to the frontal (a) per the neuraxins (n) and is frequent for emphasis. (B) Fluorescent photomicrograph of a brain slice after BrdU (bromodeoxyuridine) inclusion at 30% embryogenesis, demonstrating the bilaterally symmetrical spread of identified mitotically active neuroblasts like Z, X, Y, and W in the median PC of each brain hemisphere. (C) The lineages linked with identified neuroblasts are revealed using a 3D reconstruction derived on z-stacks of confocal images after nuclear labeling with DAPI at 35% embryogenesis (white stars). False colors are used to represent lineages (Izergina et al., 2009).

Source: https://www.researchgate.net/figure/Organization-of-neural-stem-cells-neuroblasts-in-the-early-embryonic-brain-of-the_fig1_307475381.

Each hemisphere's four neuroblasts (designated W, X, Y, and Z) perform a critical role in FB formation by establishing the neuropil's fundamental columnar arrangement. According to genetic analysis, each hemisphere of the neurodevelopment of Drosophila contains a similar set of neuroblasts, and these lineages contain the numerous tabular or small-field neurons that project to, innervate, and interconnect the PB, FB, EB, and noduli of the central complex (Bayraktar et al., 2010; Young and Armstrong, 2010a, b; Riebli et al., 2013).

Because a neuroblast consumes a stereotypic spot inside the neuroepithelium, and the lineages preserve their topological position in the CNS (Figure 3.4(C)), such cell clusters can be profiled temporally, physiologically, and biochemically (Goodman et al., 1979, 1980).

3.3.2. Modular Organization of Projections and Lineages

The central complex is related with four separate clusters of neurons in the cortical architecture of the PI area of the brain (Figure 3.5(A)). The truth that axons from each cluster form a separate tract in which they persist and subsequently project to a restricted subset of commissural fascicles of the FB confirms this relationship (Boyan et al., 1993; Boyan and Williams, 1997). The neuroblasts, their neuronal roots, and the regions they produce may thus be thought of as separate clonal or modules units, like the mechanism-building association centers in the insect (Ito et al., 1997; Lee and Luo, 2001; Ito and Awasaki, 2008).

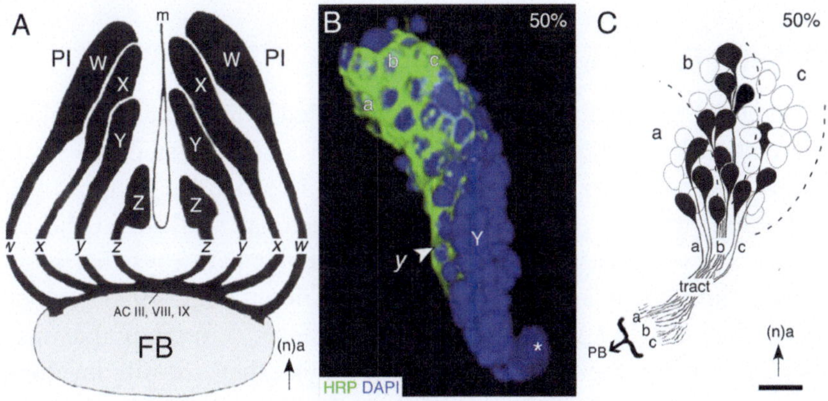

Figure 3.5. *Ancestries of embryonic progeny and the central complex's arrangement of projections into tracts.*

Note: (A) The contours of the W, X, Y, and Z lineages of progeny, as well as the related tracts of axons projecting to identify commissural fascicles of the embryonic (100%) FB, are shown in this reconstruction depending on sequential brain slices. According to the neuraxis, the arrow points to the anterior (a) (n). (B) 3D reconstruction of a typical Y (central complex lineage) at 50% embryogenesis using z-stacks of confocal images after double labeling with the neuron-specific marker HRP (green) and the nuclear stain DAPI (blue). Within the lineage, the Y neuroblast, the first y tract, and subclusters (a, b, c) of mature neurons are shown. The anterior (a) is indicated by the arrow, and it relates to all panels. (C) At 50% development, a drawing from serial sections depicts the soma axon and positions projections of individual cells from a, b, subcluster of cells at the tip of a typical central complex lineage (Z). The topological arrangement of axon projections inside the tract is determined by the position of the stomata throughout the lineage (Ito and Awasaki, 2008; Izergina et al., 2009).

Source: https://www.researchgate.net/figure/Lineages-of-embryonic-progeny-and-organization-of-projections-into-tracts-of-the_fig2_307475381.

The children of a specific lineage are formed in a temporal sequence and keep their place within the cluster according to birthday, much as they are in the ventral nerve cord, thus the lineage obtains a temporal topology. After neuron-specific labeling, reconstruction of a central intricate lineage demonstrates that early-born, mature neurons at the lineage's tip are already creating the first tract, while later-born, immature neurons closer to the neuroblast have yet to convey the label (Figure 3.5(B)). Axon tracing also demonstrates a strong link between cell body position inside a lineage and the structure of axon projections into the corresponding tract on their way to the ancient FB (Figure 3.5(C)).

3.3.3. Establishing a Neuroarchitecture

Developmental investigations have revealed that the complicated neuroarchitecture of the adult FB develops in a step-by-step way in both the grasshopper (Figure 3.6(A)–(C)) and the Drosophila (Figure 3.6(D)). The establishment of a primary orthogonal axonal scaffold in the brain is facilitated by conserved cellular and molecular pathways that may be conserved throughout species (Arendt and Nübler-Jung, 1996; Tomer et al., 2010) (Figures 3.6(A), C(i), D(i)). At this stage, commissural axons are firmly coiled within their fascicles, indicating that they are still developing. It is noteworthy that to travel the midline of the brain, the w, x, y, and z tract pioneers make use of the existing axonal scaffold that was established by the commissural pioneers earlier in the process (Williams and Boyan, 2008). Given that these y, x, w, and z tract pioneers have remained committed to their commissural fascicles, axonal remodeling must involve cells from each lineage that are either born later or create axons later in their development.

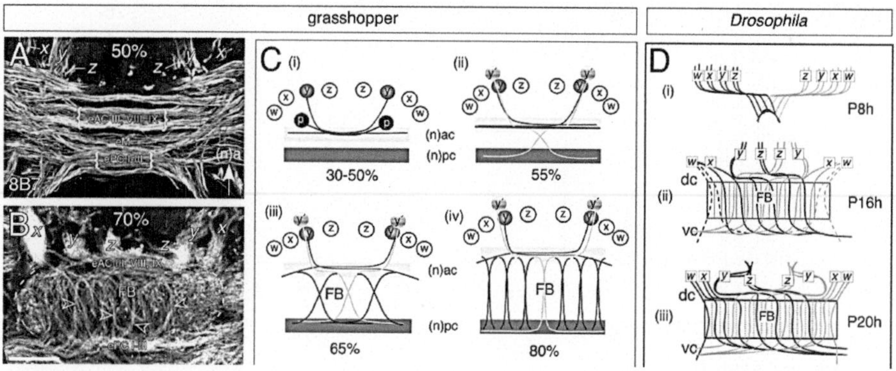

Figure 3.6. *The columnar neuroarchitecture of the FB in grasshoppers is generated by decussation of commissural axons from the W, X, Y, and Z lineages (A, B).*

Note: Axons from the x, y, and z areas project into a commissural system of axons arranged into parallel fiber bundle in a confocal view of a brain slice following 8B7 immunolabeling at 50% embryogenesis. Axon decussation has not yet happened, however fibers from the eAC (embryonic anterior commissures) III, VIII, IX, the eM (median commissure), and the embryonic posterior commissures (ePC) I-III will be involved. Panels (A–C) include an arrow that denotes anterior (a) as per the neuraxis (n). (B) Confocal picture of a brain slice following 8B7 immunolabeling at 70% embryogenesis appears axons from the x, y, and z tracts entering eAC IX, VIII, and III and subsequently decussating to ePC I-III, resulting in the FB's first fiber columns (open white arrowheads). (C) Schematics (not to scale) depicting the decussation method used to create the FB columns. The Y lineage in this case represents the overall process. (Ci) Pioneer axons (y) from the Y lineage within every brain hemisphere fasciculate with the pioneers (p) of the anterior commissures between 30 and 50% of development but remain inside their original fascicles as they cross the midline. (Cii) Later born neurons (y') from the Y lineage accompany the pioneers into the commissural system at around 55% of development, but subsequently decussate to enter the posterior commissural (pc) system. (Ciii) Axons from the other central complex lineages repeat the decussation process at 65% of development (W, X, Z). Decussation of consensually homologous axons happens at stereotypic regions around the midbrain, indicating where future columns would be located. (Civ) The midbrain neuropil expands as dendritic arborizations grow (not shown) at 80% embryogenesis, driving the pc and ac commissural fiber bundle apart and causing the columnar fiber bundles to become more orthogonal. At these ages, the neuroarchitecture of the FB by now be like that of the adult brain (see Figure 3.3(A)). (D) Decussation in Drosophila through pupal improvement follows the same pattern as in the embryonic grasshopper, as shown by projection patterns of small-field neurons summarized from anti-Echinoid labeling. After the puparium has formed, the time is stated in hours (Ph). The neurons of the right hemisphere of the brain (gray) are layered on the neurons of the left hemisphere (black). (Di) By P8 h, double fiber bundles, which we interpret as the grasshopper's w, x, y, and z tracts, had projected to the midline, and decussated to begin columnar formation of the primordial FB. (Dii) Axons are decussating at specified sites across the midline by P16 h, resulting in FB columns. (Diii) Fibers project topographically to posterior neuropils through the FB by P20 h (not shown). In Drosophila, the body axis is used to name the commissures.

Source: https://www.researchgate.net/figure/Schematics-not-to-scale-demonstrate-the-conserved-pattern-of-axon-decussation-in-the_fig1_274782193.

Topographic decussation of axons across the cerebral midline is required to change this primary orthogonal ground plan into strong chiasmal neuroarchitecture. In order to produce the columnar neuropil of the mature FB, homologous masses of later-born neurons from every protocerebrally hemisphere redirect their axonal growth cones from a frontal to a posterior commissural subsystem at stereotypic locations (Figures 3.6(B), C(ii–iv), D(ii); Boyan et al., 2008b). The columnar structure of fiber bundles inside the FB is finally hinged at the locations where de- and re-fasciculation occurs. As additional axons decussate in both systems, the columns thicken, and the distance among the commissural subsystems expands as the dendritic arbors of other innervating neurons expand. As a result, the staves become increasingly orthogonal in alignment (Boyan et al., 2015, 2008b). At 70% development, the neuroarchitecture of the grasshopper's chiasmal system matches that of an adult (see Figure 3.3(A)).

Since decussation is topographically based—axons from median lineages design the utmost across the midline, although those from more adjacent lineages project the least—the data support the hypothesis that a specific label, or a signal gradient, distributed beside the medio-lateral axis instructs neurons at specific locations inside the lineage as to where to make (see Boyan et al., 2015). Even while axogenesis occurs post embryonically in Drosophila, rather than from major embryonic inhabitants as in the grasshopper, the design of axogenesis is strikingly comparable across species, indicating that a conserved molecular signal is responsible for this phenomenon.

3.4. NEUROACTIVE SUBSTANCES AND NEUROARCHITECTURE

The FB, for example, has been shown to contain not just two systems of columnar portions, but also several layers incorporating projections from pontine cells of the tangential and PI fibers from cells somewhere else in the protocerebrum. This neuroarchitecture is mirrored through a neurochemical architecture, as seen in the axonal projections of subsections of neurons from the PI supposed to express Allatostatin (AST) and serotonin (5HT) in adults (Figure 3.7(A)), staining as a marker for NO (nitric oxide) previously at 85% embryogenesis (Figures 3.7(B)(i, ii)). Other molecules (Homberg, 2002; Herbert et al., 2010; Beetz et al., 2015) whose contributions to adult behaviors are steadily being discovered (Nässel, 2002; Winther et al., 2006; Kahsai and Winther, 2011) can also demonstrate this con how and when these stereotypic neurochemical projection patterns emerge during embryogenesis is the central topic of our investigation. We need to investigate earlier for the beginnings of the central complex's neurochemical architecture because it possesses adult traits long earlier hatching (see Boyan et al., 2015).

Neurochemicals have been found to activate neuronal networks in grasshoppers before the manifestation of adult behavior (Kutsch and Stevenson, 1986), and their production in the central complex begins through embryogenesis when the mother neuroblasts are however present (Herbert et al., 2010; Liu and Boyan, 2014). This is a significant benefit since, if the initial neuronal inhabitants can be age-profiled (see Figure 3.5(B) and (C)), expression may be traced to the temporal topology of a pedigree, allowing a biochemical "fingerprint" of the central complex to be established at various embryonic ages. This might lead to a better understanding of the developmental origins of core complex-related behaviors (Seidel and Bicker, 1996, 2002).

In this chapter, we look at a limited number of neuroactive chemicals and how they arise in the grasshopper's central complex. These are divided into three categories: (a) neuropeptides such as periviscerokinin/pyrokinin (PVK/PK), leucokinin-1 (LK-1), AST, locustatachykinin (LTK), and FLRF (FLRFamide); (b) a traditional transmitter such as 5HT; and (c) a gas such as NO. Even though not thorough, we believe that the embryonic utterance patterns of this selection of substances represent the key tendencies in the developing central complex's neurochemical and neuroanatomical architecture.

3.4.1. Locustatachykinin (LTK)

When released into the body, locstachykinin has a wide range of pathological and physiological effects, and the consequences of these effects might change significantly based on the activation of various receptor subtypes. Tachykinins have been found to be expressed at an early stage in the development of a variety of insect nervous systems (Nässel, 2002), and there are few proofs that members of the tachykinin family can act as neurotrophic factors (Satake et al., 2003), which may be like their role in the progress of craniate respiratory systems (Wong-Riley and Liu, 2005). A total of six unique kinds of LTK-immunoreactive neurons with implications in the central compartment of the grown grasshopper have been identified by Vitzthum and Homberg (1998): two peripheral neuron groups (LTT I-II), and four columnar neuron groups.

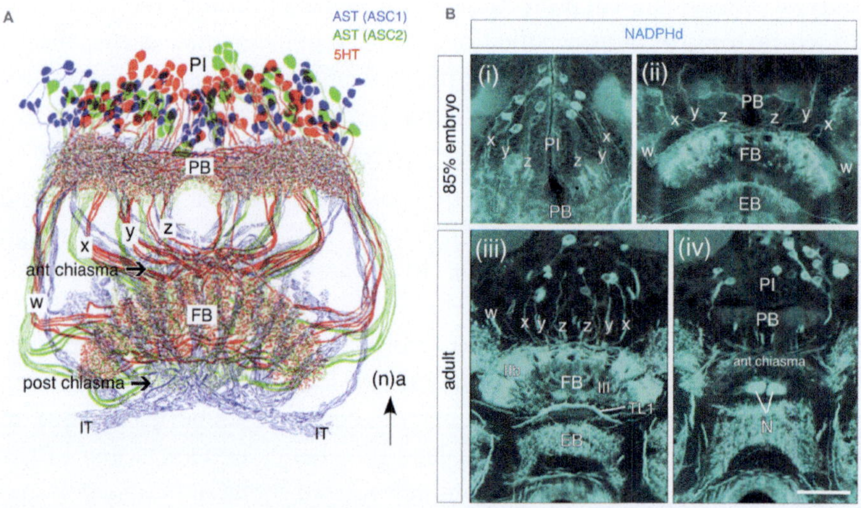

Figure 3.7. *The central complex's neurochemical architecture is based on.*

Note: The following immunolabeling against serotonin (5HT, red) and Allatostatin, reconstruction shows a very preserved sample of projections from mutual inhabitants of columnar cells in the PI to the FB through the y, x, w, and z areas. Inside the FB, axons form columnar fiber bundles, which they subsequently project to the LAL through the IT (not shown).

Source: https://www.frontiersin.org/articles/10.3389/fnbeh.2016.00167/full.

The frontal (a) as per the neuraxis (n) is shown by the arrow, which applies to all panels. (B) NO (Nitric oxide)-positive cell bodies after columnar and pontine neurons in the PI with estimates into the w, x, y, and z areas and then to the FB through the PB are visible in frontal units through with the brain of an 85% embryonic (Bi,ii) and adult grasshopper after diaphorase staining. (iii) Columnar neurite staining in layer III of the adult FB. Between the FB and the EB, tangential TL1 neurons' axonal fibers may be seen. (iv) Columnar neuron neurites extend via the PB into the frontal chiasma and subsequently to the noduli (N) according to neuraxins (Figure 3.8).

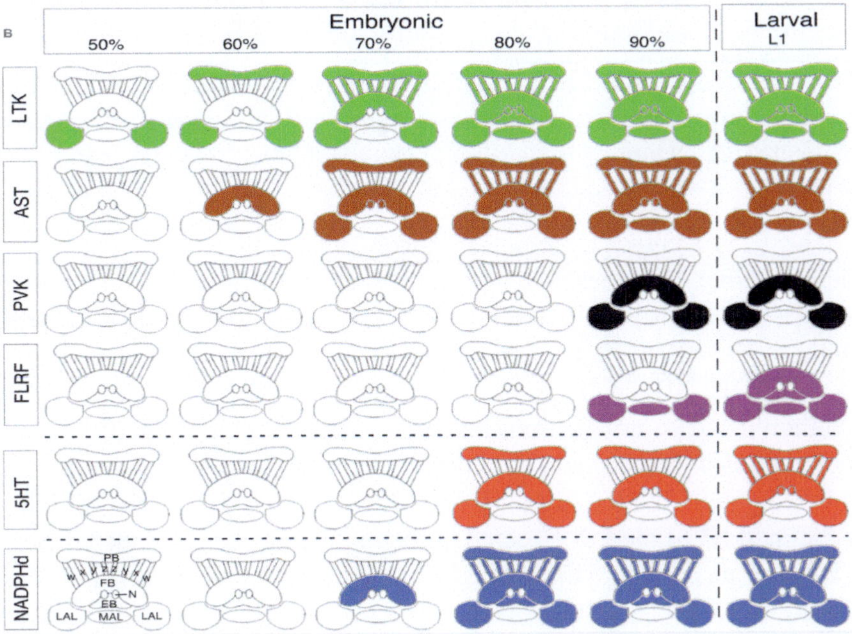

Figure 3.8. *Neurochemical expression in the central complex throughout development.*

Note: (A) Confocal pictures of LTK immunolabeling in the EB and FB at various stages of embryonic (percent) and postembryonic (larval stage 1, L1) development. LTK labeling can be seen in the early fiber columns, an upper layer of the EB, and region III of the FB at 70% embryogenesis. LTK labeling has been extended to encompass the Ib layer of the FB as well as the majority of the EB, with an overall success rate of 85%. LTK labeling intensifies in layer Ib, region III, and fiber columns of the FB at 90% of embryogenesis. The fiber column's doublet structure is visible. LTK labeling in the EB grows to levels observed post embryonically around the end of embryogenesis (98%) (L1). LTK labeling rises considerably in neuropilar subregion Ia of the FB at L1, indicating increased dendritic arborizations. All panels include an arrow that denotes anterior (a) as per the neuraxis (n). MAL is another acronym. (B) Schematics show the sequential representation of neurochemicals in the central complex from embryonic (percent) to postembryonic (L1) stages. Immunoreactivity is shown by different colors. The vertical dashed line represents hatching from the egg. Representative neuropeptides, a classical transmitter, and the endogenous messenger NO synthesizing enzyme (NADPHd) are grouped into the data. The CX's modules are illustrated in framework and represent the following, as labeled at the bottom left: PB, protocerebrally bridge; w, x, y, and z regions; fan-shaped body, FB; N, noduli; EB, LAL; ellipsoid body, median accessory lobes; MAL, lateral accessory lobes; ellipsoid body, EB; LAL, median accessory lobes; MAL, lateral fitment lobes.

Source: https://www.semanticscholar.org/paper/Development-of-the-Neurochemical-Architecture-of-Boyan-Liu/68eb9048063f4e01a7333a62c8b92129fc5aa197.

The LTC I cells are found anteriorly in the PI and project into the columnar tracts of the FB, where they give birth to arborizations in the PB, EB, and lateral bulb, among other locations (LBU). Also positioned frontal to the PB, LTC II neurons project through the columnar fibers to arborize in the LBU and PB. The columnar LTC III cell group is made up of eight big and 16 tiny neurons that are located anterior to the PB and have arborizations in layers I and II of the FB, respectively. Neurons of the LTC IV development into the posterior groove of the arborize and frontal lobe in layers I and IIa of the FB. Located in the inferior median PC, the tangential LTT I neurons project to the LAL and confound in the LBU, which is where they are named. They subsequently proceed to project through to the IT (isthmus tract) to the EB (endobronchial tract). They are behind PI neurons with extensive arborization in the PB, and they terminate in the contralateral brain's posterior median PC.

LTK immunoreactivity is identified in tangential neurons of the LALs in the embryonic central complex at 50% of embryogenesis, indicating that the central complex is already forming (Figure 3.8(B)). The terminals of these neurons are in the region of the evolving EB, which is posterior to the future FB. The pattern is consistent with the fibers belonging to the tangential LTT I neurons identified by Homberg and Vitzthum (1998) in adulthood as going to the tangential LTT I neuronal population. LTK expression begins to be expressed by an extra tangential projection system going to the LTT II group when the system reaches 60% of its development. The stained structures found in the LALs colocalize with the IT and the adult LBU, indicating that they are related. In addition, at this point, the LTK immunostaining in the PB is only moderately strong. The earliest appearance of LTK immunoreactivity in columnar neurons occurs at around 70% of embryonic development (Figure 3.8(A)). The PB, the y, x, w, and z tracts, as well as the EB, all exhibit significant immunoreactivity. The presence of LTK positive arborizations in the anterior half of layer Ib of the FB is consistent with the fact that they come from the LTC I and II neurons of the adult brain.

Immunoreactivity for LTK is detected in arborizations inside the FB, as well as in the posterior half of the moduli and the median accessory lobe (MAL), by the 85th percentile of embryogenesis. However, the intensity of this immunoreactivity increases over time (particularly in the EB), and the overall pattern remains unchanged over time. The presence of immunoreactive arborizations emanating from the LTC III columnar cell group in layers I and II of the FB, as well as LTK expression in the posterior groove, indicates the presence of LTC IV neurons in the early postembryonic phases (Rankin et al., 1998; Dircksen et al., 1999; Kreissl et al., 1999). It is already possible to see similarities between the immunoreactive design currently and that of an adult.

3.4.2. Allatostatin (AST)

The ASTs, also known as schistostatins (Schoofs et al., 1998), have previously been discovered in the adult central complex and have been shown to have a pleiotropic effect on a variety of processes. They work as neurohormones by inhibiting the production of JH (juvenile hormone) by the masses allata, and as inhibitory compounds, they are also engaged in the control of muscular contraction and the development of neuronal circuits, among other things. In *Schistocerca gregaria*, it has been demonstrated that JH regulates aggregation behavior as well as olfactory processing (Ignell et al., 2001). The presence of ASTs in the brain might therefore be compatible with their involvement in the development of synaptic networks for locomotion, antennal, and stridulation-based behaviors, all of which are known to be mediated by the central complex (Figure 3.9) (Kunst et al., 2011).

After 60–65% development, AST representation in the central complicated arises in tangential projection neurons for the first time (Figure 3.8(B)). The form and position of these initial ASTimmunoreactive fibers suggest that they are projections of AST one neurons, which is consistent with previous findings (see Figure 3.7(A)). At the present time, in the PB there is no immunoreactivity, y, x, w, or z pamphlets, or in the LALs for AST. It is reasonable to assume that the AST three neurons reported by

Vitzthum and colleagues (1996) are represented by AST-immunoreactive horizontal fibers splitting off in layers IIb and Ib.

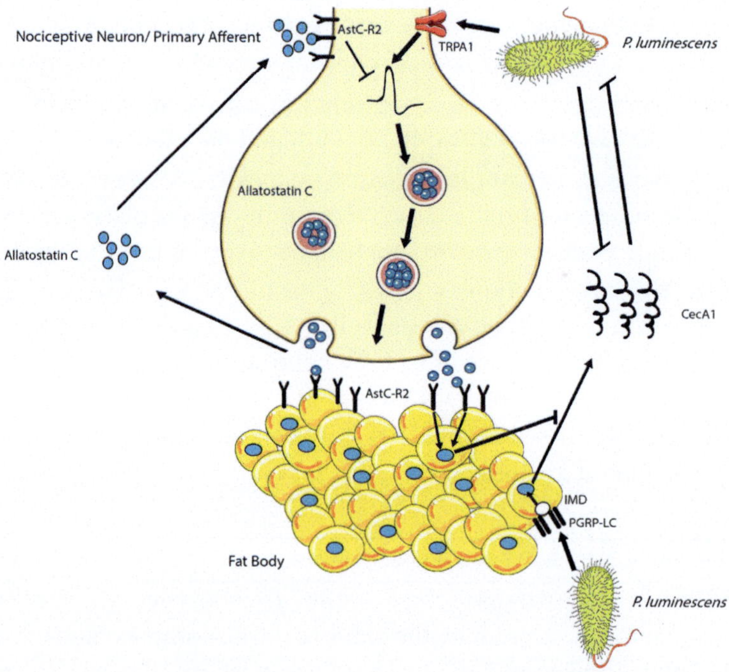

Figure 3.9. *Potential role for Allatostatin (AstC-R2) in nociceptor-bacterial-immune interactions in drosophila.*

Source: https://www.nature.com/articles/s41598-018-25855-1.

The AST two estimate system is likewise labeled in the IT, as well as in the columnar and layer IIB system of the FB, among other places (see Figure 3.7(A)). Immunoreactivity may be seen in all four of the AST tangential projection systems beginning at 80%, as well as in the columnar neurons and their methods in the prefrontal cortex (PB), the y, x, w, and z swathes, the LALs, and the moduli. It is believed that the ASC one neurons are the first to express AST in the columnar system, based on their labeling in layers Ia and Ib of the FB and their location in layers Ia and Ib of the FB. With later development, more labeling arises in the MAL and in branching twigs inside the FB, which is compatible with the presence of AST in the columnar ASC one neuron. Afterwards, the density of the AST-immunoreactive arborizations rises, and the pattern of the first instars becomes nearly identical to the pattern seen in the adult stage (Eckert et al., 1999).

3.4.3. Pyrokinins and Periviscerokinins (PK and PVK)

PKs and PVKs are the main neuroactive sections of the abdominal ganglia's neurosecretory organs, and they are also found in CNS interneurons in various insect varieties Herbert et al. (2010) found PVK/PK immunoreactivity in the peripheral system, the LAL, and all CB layers in mature grasshoppers. Immunoreactivity can also be seen in the leading peripheral method of the FB. Two immunoreactive cell groups in the inferior PC send axons into the FB through the LALs. The first group projects into layer Ib of the FB via the ipsilateral LALs, while the second group plans via the ipsilateral LAL into the contralateral LAL and interested in the FB via the contralateral LAL.

PVKs emerge in the cell group projecting through the LAL into layer Ib of the FB very late in embryogenesis (90%) (Figure 3.8(B)). However, substantial PVK staining is detected in the lateral PC and neurosecretory ways early (80%). The columnar PI neurons, PB fibers, EB, and noduli do not show any staining. Early postembryonic stages have an immunoreactive pattern that is like that of adults. PVKs are famous to have diuretic and myomodulatory effects (Wegener and Predel, 2006), therefore their postembryonic presence might be linked to olfactory cues controlling the shift from a vitellus to a vegetative diet after hatching.

3.4.4. FLRFamide

FLRFamide-like peptides are mostly found in the adult central complex's tangential projection neurons. These neurons connect into the attachment lobes, then into the EB and layer I of the FB through the IT (Homberg et al., 1999). FLRFamide-like immunoreactivity first arises in the inner complex (90%) and then in the fibers of the LALs and MALs throughout development (Figure 3.8(B)). At 65% of development, other brain areas, on the other hand, display intense FLRFamide-like immunoreactivity (Figure 3.10).

During the first instar, faint staining is observed in tangential fibers that follow columnar regions of the EB and FB, and this staining grows stronger as the larvae progress through the larval stage. Flrfamide, which is a neurohormone, is secreted by the corpus cardiaca corpora allata system and controls the heartbeat, modulates the reduction of leg muscles, and has a myoinhibitory impact on the locust fallopian tube (Robb and Evans, 1990; Peeff et al., 1994). In free-living developmental phases, these roles are more expected to be related to accumulation and the approach of sexual maturity; nevertheless, a precise role about the central complex has not yet been discovered.

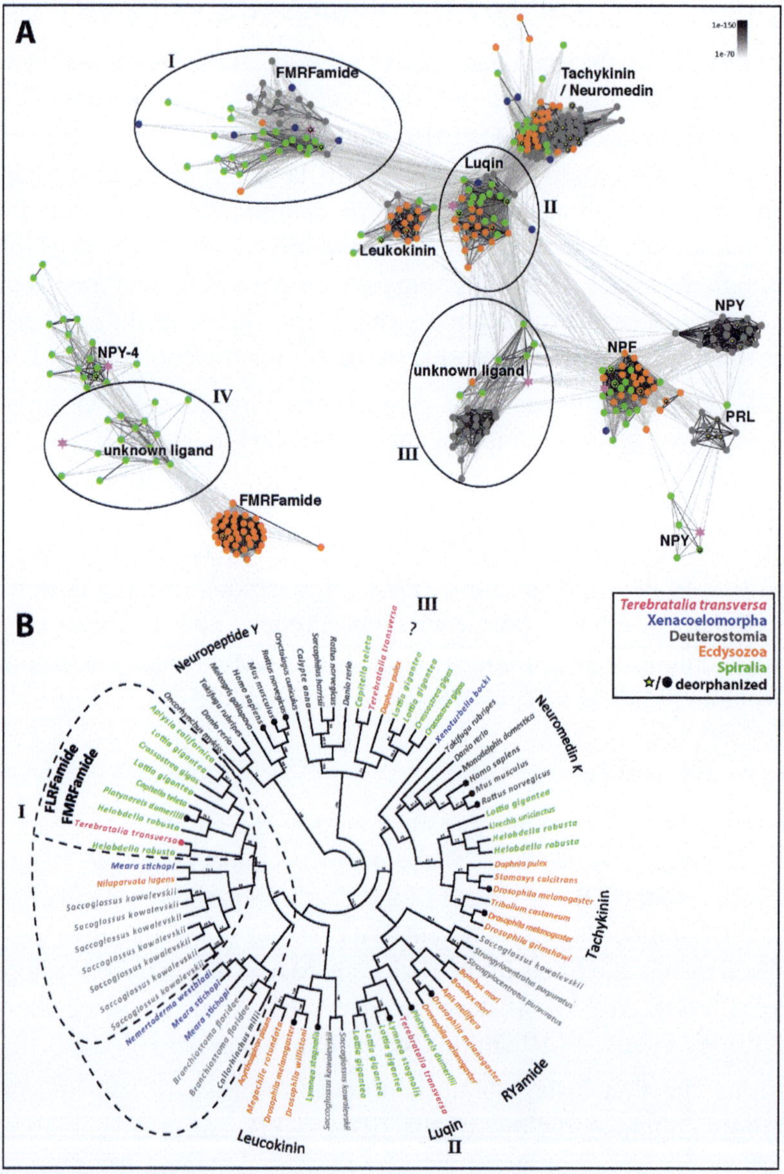

Figure 3.10. *Phylogeny of the T. transversa-FLRFamide receptor.*

Note: (A) cluster map of metazoan neuropeptide GPCRs that have been shown to be connected to FMRFamide receptors has been created. Connections with p-values ranging from 1e-70 to 1e-70 belong to BLASTP connections. The receptors I-IV, which were evaluated for activation by FLRFamide, are grouped together in a circle. NPY (neuropeptide Y), NPF (neuropeptide F), and PRL (prolactin-releasing peptide) are all neuropeptides. (B) The cladogram of neuropeptide GPCRs that revealed links to the T. transversa FLRFamide receptor in the cluster map is shown below (I). It is indicated by dashed lines which receptor groups are associated with the Terebratalia FLRFamide receptor. Receptors are shown by branches with a full circle at the end.

Source: https://www.researchgate.net/figure/Phylogeny-of-the-T-transversa-FLRFamide-receptor-A-Clustermap-of-metazoan-neuropeptide_fig3_345012535.

3.4.5. Serotonin (5HT)

The central complex of adult Schistocerca has been widely studied for 5HT immunoreactivity. The central complex's tangential serotonergic projection system is made up of five more large-field neuron pairs or groups: The inferior PC's S2 neurons project to the PB and derriere optic tubercules; the inferior PC's S3 neurons run over the LAL and into layer Ia of the FB; the fronto-median PC's S4 neuron pair lends credence to varicose local divisions in layer Ib of the FB; the posterior PI nerves' S5 neuron pair layer Ia of the FB; and the S6 pair developments into the posterolateral LALs where (Figure 3.11).

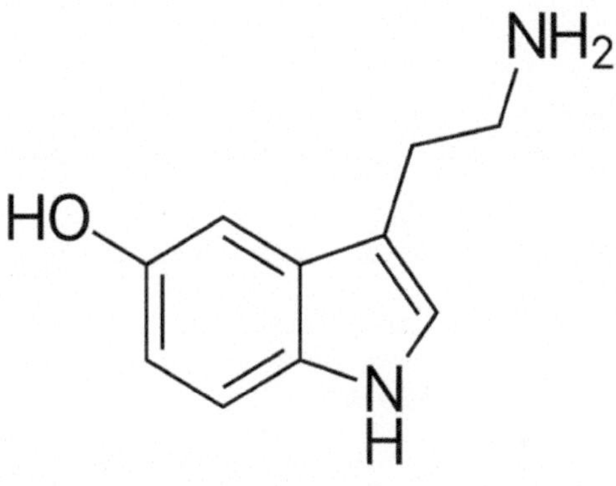

Figure 3.11. *Structure of serotonin.*

Source: https://en.wikipedia.org/wiki/Serotonin.

In the grasshopper, considerable 5HT immunoreactivity is detected in the central complex just after 75% of embryogenesis, but it increases dramatically throughout subsequent development and becomes highly detectable (Figure 3.8(B)). 5HT emerges earliest (75–80%) in peripheral projections, in a way comparable to that of members of the AST and LTK peptide families and may be found in the columniform system just before hatching, according to the results of the study. Tangential projection neurons begin to describe 5HT in their predictions to the PB (S2) and into the FB via the LALs at the 80% stage (S5). Additionally, there are serotonergic fibers in all three S3 and S4 groups, demonstrating that 5HT is represented in all three S4 and S3 groups. Every single one of these serotonergic tangential bulge neurons is immunoreactive at the same time. 5HT immunoreactivity, on the other hand, has not yet been discovered in the columnar w, x, y, and z tracts, as well as in PI neurons. The intensity of staining in the tangential projection neurons increases in the late embryo (90–95%) when the neuropilar volume of the central complex develops because of increased fiber density from extra in growing neurons. The presence of 5HT immunoreactivity in the columniform system can be spotted for the first time at 99%, just already hatching, and the complete pattern is very like that of adult stages and the L1, even though the complete list of S1 immunoreactive cells is not obvious now.

In adult bugs, 5HT regulates circadian rhythms (Sitaraman et al., 2008; Zars, 2009), is required for spatial learning and memory (Saifulla and Tomioka, 2002; Yuan et al., 2005), modulates aggression, and mediates the phase change from solitary to gregarious. Gammie and Truman (1997) discovered that neuroactive chemicals are hierarchically arranged, and 5HT has been proposed as an overall coordinator of axogenesis, neurogenesis, and biochemical and cellular variation during development. Chemical signals are required for serotonergic neurons to become functionally active (Condron, 1999), together with 5HT (Sykes and Condron, 2005), implying that both feedback and feedforward systems are engaged in the improving grasshopper nervous system.

3.4.6. NADPHd (Nitric Oxide)

The central complex and adult MB have NADPHd activity (O'Shea et al., 1998; Kurylas et al., 2005; Siegl et al., 2009). In developing insect nervous systems, NO has been related to retinal patterning, cell proliferation, neuronal maturation, and axogenesis, synaptogenesis (Kuzin et al., 1996; Truman et al., 1996a, b). Kurylas et al. (2005) report a thorough examination of NO immunolabeling related with the central compound of the adult grasshopper (Figure 3.12).

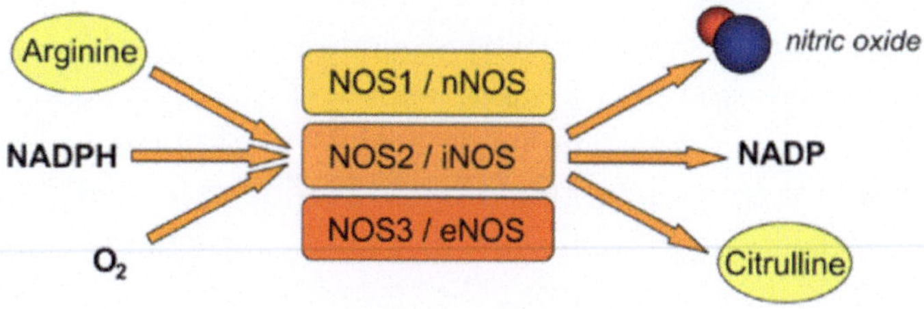

Figure 3.12. *Synthesis of nitric oxide.*

Source: https://www.reading.ac.uk/nitricoxide/intro/no/synthesis.htm.

Six neuron types with tangential, pontine, and columniform projection systems were found (approximately 170 neurons): Columnar neurons (a) The columnlike NO system of the FB is supported by around 50 neurons after the latter PI. Their fibers go from the PB to layer III of the FB and then to the moduli through the w, x, y, and z tracts. Tangential neurons. Their somata (about 20) are positioned behind the PB. Six bilateral sets of neurons make up the fourth tangential neuron group (TL4), which runs down the w tract to the LALs. Layer II of the FB also contains terminal arborizations. The fifth peripheral arborization system is found in the anterior lip, rump commissure I, and layer II of the FB. (c) Neurons in the pontine ganglion. As per neuraxis (see Figures 3.13(B)iii, iv), these neurons transmit via the PB to the FB through the anterior chiasm and provide to layer I and layer IIb columns in the FB. They also link the columns of these strata together.

Whereas in the tangential systems NADPHd activity grows in a piecemeal way as more fibers innervate the area, it occurs abruptly and simultaneously in all parts of the columnar system of the FB, suggesting a temporally organized method for generating this neurochemical building (Williams et al., 2005; Herbert et al., 2010).

3.4.7. Interim Summary

The postnatal expression design of neurochemicals in the central complexity has several regular properties. To begin with, all the chemicals studied are expressed in the embryonic central complex modules in a substance-specific temporal pattern rather than synchronously. Specific modules developed immunoreactive only after hatching for some, whereas others appear to have finished their evolving plan throughout embryogenesis. Second, when neuropilar modules are considered, immunoreactivity emerges early in the FB and LAL, and later in the PB and x, w, y, and z areas. Finally, a few neuropeptides are expressed sooner. If we merely look at the columniform system, we see a like pattern: LTK-related peptides come first (70%), followed by diaphorase and AST at 75–80%, and 5HT just post embryonically. Finally, during development, immunoreactivity does not arise in an all-or-nothing fashion; there is a definite progressive increase in signal intensity for each neurochemical. We can show that this is attributable to an increase in the number of immunoreactive cells from a specific lineage conveying the appropriate neurochemical (see Figure 3.13(D)).

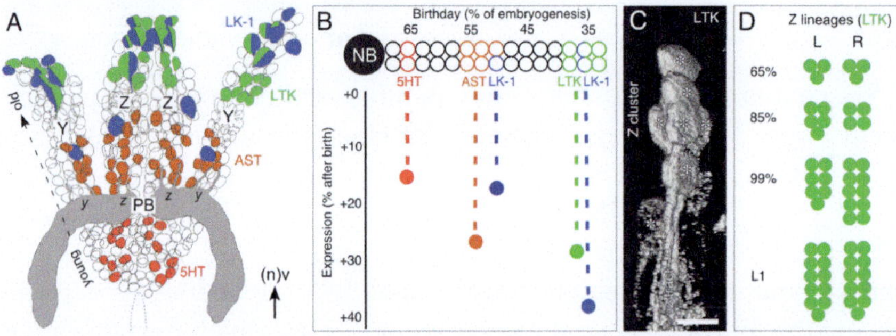

Figure 3.13. *Topology of neurons expressing various neurochemicals in central complex lineages.*

Note: (A) Serial sections were used to reconstruct this image, which reviews the location of suggest implications the neurochemicals AST (brown), 5HT (red), leucocidin, and LTK inside the bilateral Y and Z lineages at 100% of embryogenesis. The image was created using immunolabeling. Several neurons convey the neurochemicals LTK and LK-1 in tandem with one another. When it comes to their lineage, neurons that express the numerous neurochemicals are in stereotypic sites (which also reflect age): the youngest (newly born) cells are found at their base, and the eldest (oldest born) cells are found at their tip. Tract doublets (see Figure 3.3(C)) are used by neurons to route axons to the PB. At this age, neuroblasts are no longer present in the body (Boyan and Liu, 2014). According to the neuraxins, the arrow represents the anterior (a). (B) The temporal expression of neurochemicals 5HT, AST, LK-1, and LTK from a typical central complex lineage is depicted in this schematic. Neurons occupy stereotypical places within the lineage that correspond to the date of their birth. The temporal interval between birth and the onset of neurochemical expression is represented by the vertical axis. Even while LK-1 cells at the tip of the lineage are born very early in embryogenesis, they remain immunohistochemically inactive for over 40% of the whole embryonic development. 5HT cells near the base of the lineage are born late, but they produce their neurochemicals within a short period of time after they are born. It is predicted from the findings that neurochemicals should manifest themselves in a progressive manner inside the central complex neuropil. (C) A bunch of LTK-positive cells near the top of the Z lineage, together with their predictions creating the z tract, can be seen in this 3D picture from confocal z-stacks subsequent immunolabeling at 80% of embryogenesis. (D) The number of LTK-immunoreactive cells in the left (L) and correct (R) Z lineages at different stages of development was compiled from three preparations in which both Z lineages were completely visible. At 65% improvement, LTK-immunoreactive cells composed 2% of the Z lineage, 3% at 85%, 5% at 99%, and 8% at L1 embryogenesis.

Source: https://pubmed.ncbi.nlm.nih.gov/27630548/.

3.5. BIOCHEMICAL PROFILING OF CENTRAL COMPLICATED LINEAGES

Considering that the pattern in which discrete neuroactive substances seem in the adult nervous system during advancement is perilous for the adaptive functioning of the adult nervous system (Nässel, 2002), understanding about whether this neurochemical structure can be linked the lineages of central complex neurons could offer valuable insights into the behavioral role played by the central complex at various critical stages like pupation, molting, and hatching (Taghert and Goodman, 1984).

Z and Y lineages, as illustrative of central complex lineages, have been revealed to be jointly symmetrical, and because they recollect their interior organization capable of and outside hatching, they are solvable down to the level of single, distinguishable, seemingly homologous cells, according to the findings of the reconstructions (Boyan et al., 2010a). Therefore, it is possible to demonstrate that the lineages have a temporal topology, in which the placement within the lineage corresponds to the birth date of a certain cell is determined. The expression of various neurochemicals may therefore be age-profiled in a lineage (Figure 3.8; Boyan et al., 2008a, 2010a), which can then be compared to the expression designs of the similar substances in the central compound (Figure 3.8).

These analyzes show several things, including the fact that all distinct neuroactive chemicals do certainly co-localize to the similar lineage, indicating that the neuroblasts relevant for every lineage are biochemically multipotent. Second, the pedigrees are virtually equally zoned in terms of where the neurons that express these chemicals are situated, indicating that they are part of a shared developmental pathway, as previously stated.

Neuroactive chemicals have been seen to co-localize with one another at the single-cell level in some situations. The coexpression of neuropeptides in insect nervous systems has been extensively documented (Taghert and Truman, 1982a, b).

As demonstrated in Drosophila melanogaster, consecutive neuroblast divisions produce biochemically different cells, and this is supported by the unambiguous zonation of expression. The technique may include the neuroblast producing a transcript factor that is particular to a particular mitotic division or series of divisions, so offering each subsequent daughter cell with a distinct character after the first division (Liu and Boyan, 2014).

When the central compound of a hemimetabolous insect like the grasshopper hatches, all the neuroblasts responsible for the generation of neurons in the central complex die among 70 and 75% of the way through embryogenesis, indicating that the central compound of the insect is structurally complete. On a biochemical level, however, this is clearly not the situation. They are all born during embryogenesis, yet they all stay biochemically inactive for variable amounts of time afterward, as detailed in this chapter (Figure 3.13(B)).

For each material, this inactive time is stereotypical, and it may persist for more than 40% of the embryonic process. LTK-stating cells at the apical tip of the lineage, for instance, are formed soon in embryogenesis, but they remain immunocytochemically inactive for about 30% of the embryonic development. Even while the LK-1-expressing cells, which are likewise present at the apical tip of the lineage, are formed within the same time frame as the LTK-expressing cells, they produce the neurochemical only around 40% shortly, i.e., long after the LTK-expressing cells (Thompson and Siegler, 1991; Tettamanti et al., 1997).

According to what we previously found (in the section titled "Developmental Expression Patterns"), immunoreactivity in the FB throughout development does not manifest itself in a binary fashion; rather, there is a distinct stepwise rise in signal intensity for individually neurochemical through progress. At the very least, in the situation of the columnlike system, this is most likely owing to an increase in the proportion of immunoreactive cells from a specific cell mass of the PI that expresses the significant neurochemical, as opposed to a decrease (Figure 3.13(C) and (D)). When the cell cluster reaches 75%, it is no longer possible for it to grow since all the proliferative cells that are liable for these central compound lineages, comprising the neuroblasts, the tumor mother cells, and the middle progenitors, have stopped mitotically reproducing by that time (Boyan et al., 2010b; Boyan and Liu, 2014).

3.6. LINEAGE TOPOLOGY TRANSLATES INTO NEUROARCHITECTURE

To resolve the embryogenesis of biochemical expression in the central complex, we may develop a timeline utilizing the temporal and biochemical profile of a lineage, and the famous times when neurochemicals are produced in the central complex. It is now possible to create an idea for how this biochemical and temporal structure could translate into the neurochemical structure of the central compound (Figure 3.14) (Turlejski, 1996; Triphan et al., 2010).

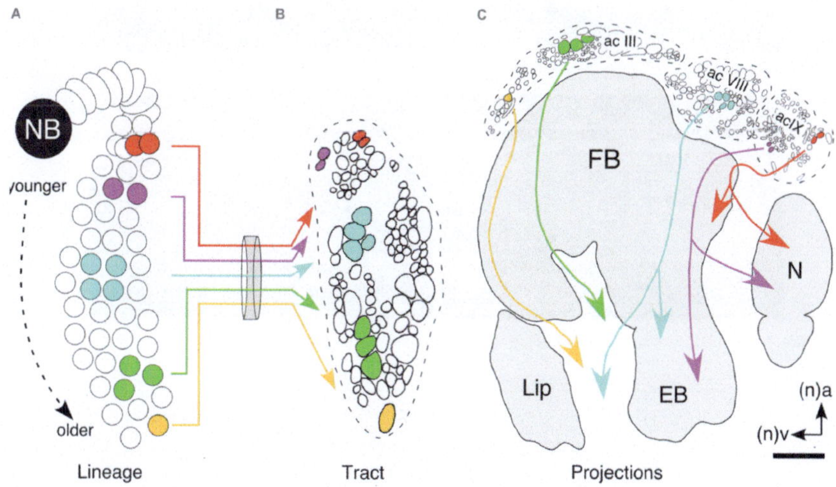

Figure 3.14. *Hypothesis on how the topology of axial projections in the FB can be translated from neuronal topology in a lineage.*

Note: (A) Neurons in this hypothetical lineage of the central complex occupy stereotypic positions based on their birthday. Colors are used to represent the expression of several neurochemicals that are not described. (B) Axon forecasts are topologically sorted as per the placement of the somata inside the lineage in a cross-section of the tract. (C) The position of axons inside a tract corresponds to their placement in the central complex's commissural fascicules (AC III, VIII, IX). These fascicles' axons decussate, resulting in a columnar neuroarchitecture in the growing FB. The neurochemical architecture of the central complex reflects the lineage's temporal and biochemical structure in this way. According to the neuraxis, arrows denote ventral (v) and anterior (a).

Source: https://pubmed.ncbi.nlm.nih.gov/27630548/.

In the hypothesized lineage of the central complex, neurons occupy stereotypic positions based on their birthday. A cross-section of the territory reveals that axon estimates are topologically organized as per the position of the somata inside the lineage, and that axon view inside (a) area corresponds to placement in the central complex's commissural fascicles (Figure 3.15) (Vitzthum et al., 1996; Vitalis et al., 2007).

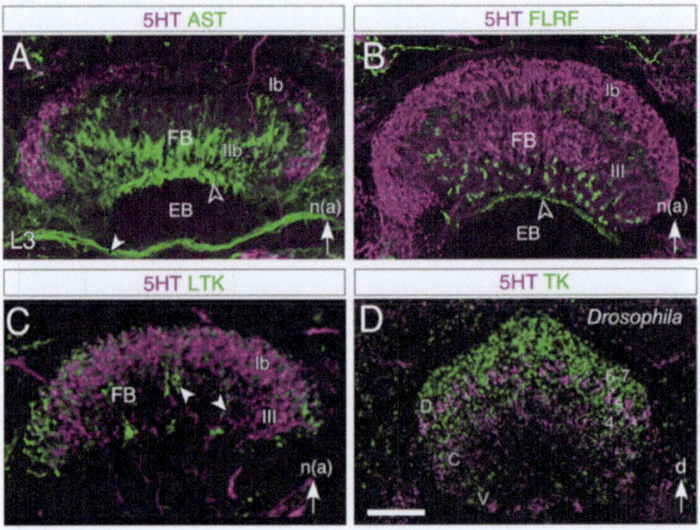

Figure 3.15. *Layered expression of neurochemicals in the FB of the central complex in the larval (L3) grasshopper (A–C) and adult drosophila (D).*

Note: (A) A schematic displays the arrangement of core enough in the brain in sagittal perspective. The letters MB stand for mushroom body, mOc for median ocellus, FB for fan-shaped body, N for nodulus, EB for ellipsoid body, and AL for antennal lobe. Arrows point to posterior (a) and ventral (v) on the neuraxins (n). Axes have an impact on all panels (Westbrook and Bollenbacher, 1990; Wegerhoff et al., 1996).

Source: https://www.readcube.com/articles/10.3389%2Ffnbeh.2016.00167.

Suppose our theory that distinct expression zones inside the W, X, Y, and Z lineages transfer into biochemical neuroarchitecture is valid. In that case, the zoning of neurochemical prediction systems in the FB might take the form of a layered expression pattern. In fact, double immunolabeling tests in growing grasshoppers expose no co-expression of 5HT and the neuropeptides

(Dip)-AST I, LK-1, FLRFamide, LTK, and PVK/PK in arborizations among these subregions of the FB (Williams et al., 2005; Herbert et al., 2010) or adult Drosophila. Because LTK and 5HT are communicated by neurons of various ages, the cellular neuroarchitecture of the FB depending on histology may now be examined from a temporal perspective via its neurochemical architecture. The ventral portions of the PB, ventral commissures, and noduli are largely serotonergic late in embryogenesis, suggesting that the axon processes come from younger cells. In contrast, the embryonic EB and columnar projections in the FB are virtually entirely LTK-positive, and hence include axons from earlier cells. This corresponds to the sequence in which core complex modules' neuroarchitecture develops (Figure 3.16) (Zitnan et al., 1999; Wolff et al., 2015).

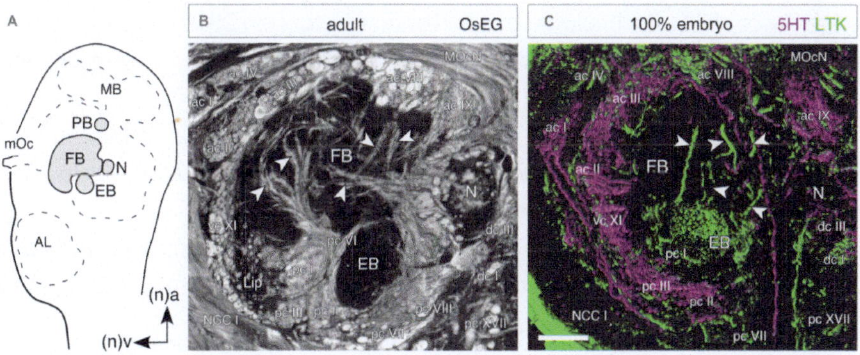

Figure 3.16. *The architecture of the central complex is mirrored in the topography of neurochemical expression across time.*

Note: (A) In sagittal view, a schematic depicts the placement of core development modules in the brain. MB stands for Borgeon body; mOc stands for median stemma; FB stands for fan-shaped body; N stands for nodulus; EB stands for ellipsoidal body; AL stands for antennary lobe. According to the neuraxins, arrows point to ventral (v) and anterior (a). All panels are affected by axes. (B) Axon profiles in posterior (pc), anterior (ac), dorsal (dc), ventral (vc) and commissures confining the N, EB, FB, and ventral lip are seen in a histological slice of the central compound in the adult brain after osmium ethyl gallate staining. The MocN and nervus corporis cardiac I are two more fiber tracts seen (NCC I). (C) Confocal picture of a parasagittal slice of the central complex at 100% development following double tagging with LTK (green) and 5HT (magenta).

Source: https://www.readcube.com/articles/10.3389%2Ffnbeh.2016.00167.

REFERENCES

1. Arendt, D., & Nübler-Jung, K., (1996). Common ground plans in early development in mice and flies. *Bioessays, 18*, 255–259. doi: 10.1002/bies.950180314.

2. Bayraktar, O. A., Boone, J. Q., Drummond, M. L., & Doe, C. Q., (2010). *Drosophila* type II neuroblast lineages keep Prospero levels low to generate large clones that contribute to the adult brain central complex. *Neural Dev., 5*, 26. doi: 10.1186/1749-8104-5-26.

3. Beetz, M. J., El Jundi, B., Heinze, S., & Homberg, U., (2015). Topographic organization and possible function of the posterior optic tubercles in the brain of the desert locust *Schistocerca gregaria. J. Comp. Neurol., 523*, 1589–1607. doi: 10.1002/cne.23736.

4. Bender, J. A., Pollack, A. J., & Ritzmann, R. E., (2010). Neural activity in the central complex of the insect brain is linked to locomotor changes. *Curr. Biol., 20*, 921–926. doi: 10.1016/j.cub.2010.03.054.

5. Boyan, G. S., & Williams, J. L. D., (1997). Embryonic development of the pars intercerebralis/central complex of the grasshopper. *Dev. Genes Evol., 207*, 317–329. doi: 10.1007/s004270050119.

6. Boyan, G. S., Williams, J. L. D., & Herbert, Z., (2008a). An ontogenetic analysis of locustatachykinin-like expression in the central complex of the grasshopper *Schistocerca gregaria. Arthropod Struct. Dev., 37*, 480–491. doi: 10.1016/j.asd. 2008.04.002.

7. Boyan, G. S., Williams, J. L. D., & Herbert, Z., (2008b). Fascicle switching generates a chiasmal neuroarchitecture in the embryonic central body of the grasshopper *Schistocerca gregaria. Arthropod Struct. Dev., 37*, 539–544. doi: 10. 1016/j.asd.2008.07.005.

8. Boyan, G. S., Williams, J. L. D., & Meier, T., (1993). Organization of the commissural fibers in the adult brain of the locust. *J. Comp. Neurol., 332*, 358–377. doi: 10.1002/cne.903320308.

9. Boyan, G. S., Williams, L., Legl, A., & Herbert, Z., (2010b). Proliferative cell types in embryonic lineages of the central complex of the grasshopper *Schistocerca gregaria. Cell Tissue Res., 341*, 259–277. doi: 10.1007/s00441-0100992-6.

10. Boyan, G., & Liu, Y., (2014). Timelines in the insect brain: Fates of identified neural stem cells generating the central complex in the grasshopper *Schistocerca gregaria. Dev. Genes Evol., 224*, 37–51. doi: 10.1007/s00427-0130462-8.

11. Boyan, G., Williams, L., & Herbert, Z., (2010a). Multipotent neuroblasts

generate a biochemical neuroarchitecture in the central complex of the grasshopper *Schistocerca gregaria. Cell Tissue Res., 340*, 13–28. doi: 10.1007/s00441-0090922-7.

12. Boyan, G., Williams, L., & Liu, Y., (2015). Conserved patterns of axogenesis in the panarthropod brain. *Arthropod Struct. Dev., 44*, 101–112. doi: 10.1016/j.asd. 2014.11.003.
13. Dircksen, H., Skiebe, P., Abel, B., Agricola, H., Buchner, K., Muren, J. E., et al., (1999). Structure, distribution and biological activity of novel members of the allatostatin family in the crayfish *Orconectes limosus. Peptides, 20*, 695–712.
14. Eckert, M., Predel, R., & Gundel, M., (1999). Periviscerokinin-like immunoreactivity in the nervous system of the American cockroach. *Cell Tissue Res., 295*, 159–170. doi: 10.1007/s004410051222.
15. Farris, S. M., & Sinakevitch, I., (2003). Development and evolution of the insect mushroom bodies: Towards the understanding of conserved developmental mechanisms in a higher brain center. *Arthropod Struct. Dev., 32*, 79–101. doi: 10. 1016/s1467-8039(03)00009-4.
16. Gammie, S. C., & Truman, J. W., (1997). Neuropeptide hierarchies and the activation of sequential motor behaviors in the hawkmoth *Manduca sexta. J. Neurosci., 17*, 4389–4397.
17. Goodman, C. S., O'Shea, M., McCaman, R., & Spitzer, N. C., (1979). Embryonic development of identified neurons: Temporal pattern of morphological and biochemical differentiation. *Science, 2014*, 1219–1222. doi: 10.1126/science. 36661.
18. Goodman, C. S., Pearson, K. G., & Spitzer, N. C., (1980). Electrical excitability: A spectrum of properties in the progeny of a single embryonic neuroblast. *Proc. Natl. Acad. Sci. U S A, 77*, 1676–1680. doi: 10.1073/pnas.77. 3.1676.
19. Hanesch, U., Fischbach, K. F., & Heisenberg, M., (1989). Neuronal architecture of the central complex in *Drosophila melanogaster. Cell Tissue Res., 257*, 343–366. doi: 10.1007/BF00261838.
20. Harley, C. M., & Ritzmann, R. E., (2010). Electrolytic lesions within central complex neuropils of the cockroach brain affect the negotiation of barriers. *J. Exp. Biol., 213*, 2851–2864. doi: 10.1242/jeb.042499.
21. Heinze, S., & Homberg, U., (2008). Neuroarchitecture of the central complex of the desert locust: Intrinsic and columnar neurons. *J. Comp. Neurol., 511*, 454–478. doi: 10.1002/cne.21842.
22. Herbert, Z., Rauser, S., Williams, L., Kapan, N., Güntner, M., Walch, A., et al., (2010). Developmental expression of neuromodulators in the

central complex of the grasshopper *Schistocerca gregaria*. *J. Morphol., 271*, 1509–1526. doi: 10. 1002/jmor.10895.

23. Homberg, U., (1991). Neuroarchitecture of the central complex in the brain of the locust *Schistocerca gregaria and S. americana* as revealed by serotonin immunocytochemistry. *J. Comp. Neurol., 303*, 245–254. doi: 10.1002/cne. 903030207.

24. Homberg, U., (2002). Neurotransmitters and neuropeptides in the brain of the locust. *Microsc. Res. Tech., 56*, 189–209. doi: 10.1002/jemt. 10024.

25. Homberg, U., Heinze, S., Pfeiffer, K., Kinoshita, M., & El Jundi, B., (2011). Central neural coding of sky polarization in insects. *Philos. Trans. R. Soc. Lond. B Biol. Sci., 366*, 680–687. doi: 10.1098/rstb.2010.0199.

26. Homberg, U., Vitzthum, H., Müller, M., & Binkle, U., (1999). Immunocytochemistry of GABA in the central complex of the locust *Schistocerca gregaria*: Identification of immunoreactive neurons and colocalization with neuropeptides. *J. Comp. Neurol., 409*, 495–507. doi: 10. 1002/(SICI)1096-9861(19990705)409, 3<495::AID-CNE12>3.0.CO;2-F.

27. Huber, F. (1960). Investigations into the function of the central nervous system and in particular the brain in locomotion and the production of sounds in crickets. *Journal of Comparative Physiology, 44* (1), 60-132.

28. Ignell, R., Couillaud, F., & Anton, S., (2001). Juvenile-hormone-mediated plasticity of aggregation behaviour and olfactory processing in adult desert locusts. *J. Exp. Biol., 204*, 249–259.

29. Ilius, M., Wolf, R., & Heisenberg, M., (1994). The central complex of *Drosophila melanogaster* is involved in flight control: Studies on mutants and mosaics of the gene ellipsoid body open. *J. Neurogenet., 9*, 189–206. doi: 10. 3109/01677069409167279.

30. Ito, K., & Awasaki, T., (2008). "Clonal unit architecture of the adult fly brain." In: Technau, G. M., (ed.), *Brain Development in Drosophila Melanogaster* (pp. 137–158). New York, NY: Springe.

31. Ito, K., Awano, W., Suzuki, K., Hiromi, Y., & Yamamoto, D., (1997). The *Drosophila* mushroom body is a quadruple structure of clonal units each of which contains a virtually identical set of neurones and glial cells. *Development, 124*, 761–771.

32. Izergina, N., Balmer, J., Bello, B., & Reichert, H., (2009). Postembryonic development of transit-amplifying neuroblast lineages in the *Drosophila* brain. *Neural Dev., 4*, 44. doi: 10.1186/1749-8104-4-44.

33. Kahsai, L., & Winther, A. M. E., (2011). Chemical neuroanatomy of the *Drosophila* central complex: Distribution of multiple neuropeptides

in relation to neurotransmitters. *J. Comp. Neurol., 519*, 290–315. doi: 10.1002/cne.22520.

34. Kreissl, S., Weiss, T., Djokaj, S., Balezinam, O., & Rathmayer, W., (1999). Allatostatin modulates skeletal muscle performance in crustaceans through pre- and postsynaptic effects. *Eur. J. Neurosci., 11*, 2519–2530. doi: 10.1046/j. 1460-9568.1999.00674.x.
35. Kunst, M., Pförtner, R., Aschenbrenner, K., & Heinrich, R., (2011). Neurochemical architecture of the central complex related to its function in the control of grasshopper acoustic communication. *PLoS One, 6*, e25613. doi: 10.1371/journal.pone.0025613.
36. Kurylas, A. E., Ott, S. R., Schachtner, J., Elphick, M. R., Williams, L., & Homberg, U., (2005). Localization of nitric oxide synthase in the central complex and surrounding midbrain neuropils of the locust *Schistocerca gregaria*. *J. Comp. Neurol., 484*, 206–223. doi: 10.1002/cne.20467.
37. Kuzin, B., Roberts, I., Peunova, N., & Enikolopov, G., (1996). Nitric oxide regulates cell proliferation during *Drosophila* development. *Cell, 87*, 639–649. doi: 10.1016/s0092-8674(00)81384-7.
38. Lee, T., & Luo, L., (2001). Mosaic analysis with a repressible cell marker (MARCM) for *Drosophila* neural development. *Trends Neurosci., 24*, 251–254. doi: 10.1016/s0166-2236(00)01791-4.
39. Liu, G., Seiler, H., Wen, A., Zars, T., Ito, K., Wolf, M., et al., (2006). Distinct memory traces for two visual features in the *Drosophila* brain. *Nature, 439*, 551–556. doi: 10.1038/nature04381.
40. Meinertzhagen, I. A., Emsley, J. G., & Sun, X. J., (1998). Developmental anatomy of the *Drosophila* brain: Neuroanatomy is gene expression. *J. Comp. Neurol., 402*, 1–9. doi: 10.1002/(SICI)1096-9861(19981207)402:1<1::AID-CNE1> 3.0.CO;2-Q.
41. Mizunami, M., Iwasaki, M., Nishikawa, M., & Okada, R., (1997). Modular structures in the mushroom body of the cockroach. *Neurosci. Lett., 229*, 153–156. doi: 10.1016/s0304-3940(97)00438-2.
42. Müller, M., Homberg, U., & Kühn, A., (1997). Neuroarchitecture of the lower division of the central body in the brain of the locust (*Schistocerca gregaria*). *Cell Tissue Res., 288*, 159–176. doi: 10.1007/s004410050803.
43. Nässel, D. R., (2002). Neuropeptides in the nervous system of *Drosophila* and other insects: Multiple roles as neuromodulators and neurohormones. *Prog. Neurobiol., 68*, 1–84. doi: 10.1016/s0301-0082(02) 00057-6.
44. Neuser, K., Triphan, T., Mronz, M., Poeck, B., & Strauss, R., (2008). Analysis of a spatial orientation memory in *Drosophila*. *Nature, 453*, 1244–1247. doi: 10. 1038/nature07003.

45. O'Shea, M., Colbert, R., Williams, L., & Dunn, S., (1998). Nitric oxide compartments in the mushroom bodies of the locust brain. *Neuroreport, 9*, 333–336. doi: 10.1097/00001756-199801260-00028.

46. Pan, Y., Zhou, Y., Guo, C., Gong, H., Gong, Z., & Liu, L., (2009). Differential roles of the fan-shaped body and the ellipsoid body in *Drosophila* visual pattern memory. *Learn. Mem., 16*, 289–295. doi: 10.1101/lm.1331809.

47. Peeff, N. M., Orchard, I., & Lange, A. B., (1994). Isolation, sequence and bioactivity of PDVDHVFLRFamide and ADVGHVFLRFamide peptides from the locust central nervous system. *Peptides, 15*, 387–392. doi: 10.1016/01969781(94)90193-7.

48. Pfeiffer, K., & Homberg, U., (2014). Organization and functional roles of the central complex in the insect brain. *Annu. Rev. Entomol., 59*, 165–184. doi: 10. 1146/annurev-ento-011613-162031.

49. Rankin, S. M., Stay, B., Chan, K., & Jackson, E. S., (1998). Cockroach allatostatin-immunoreactive neurons and effects of cockroach allatostatin in earwigs. *Arch. Insect Biochem. Physiol., 38*, 155–165. doi: 10.1002/(SICI)15206327(1998)38:4<155::AID-ARCH1>3.0.CO;2-Q.

50. Reichert, H., & Boyan, G. S., (1997). Building a brain: Insights from insects. *Trends Neurosci., 20*, 258–264.

51. Renn, S. C. N., Armstrong, J. D., Yang, M., Wang, Z., An, X., Kaiser, K., et al., (1999). Genetic analysis of the *Drosophila* ellipsoid body neuropil: Organization and development of the central complex. *J. Neurobiol., 41*, 189–207.

52. Riebli, N., Viktorin, G., & Reichert, H., (2013). Early-born neurons in type II neuroblast lineages establish a larval primordium and integrate into adult circuitry during central complex development in *Drosophila*. *Neural Dev., 8*, 6.

53. Robb, S., & Evans, P. D., (1990). FMRFamide-like peptides in the locust: Distribution, partial characterization and bioactivity. *J. Exp. Biol., 149*, 335–360.

54. Saifulla, A. S. M., & Tomioka, K., (2002). Serotonin sets the day state in the neurons that control coupling between the optic lobe circadian pacemakers in the cricket *Gryllus bimaculatus*. *J. Exp. Biol., 205*, 1305–1314.

55. Satake, H., Kawada, T., Nomoto, K., & Minakata, H., (2003). Insight into tachykinin-related peptides, their receptors and invertebrate tachykinins: A review. *Zoolog. Sci., 20*, 533–549. doi: 10.2108/zsj.20.533.

56. Schoofs, L., Janssen, I., Veelaert, D., Vanden, B. J., Tobe, S. S., & De Loof, A., (1998). Ecdysiostatins and allatostatins in *Schistocerca gregaria*.

Proc. Ann. N Y Acad. Sci., 15, 301–305. doi: 10.1111/j.1749-6632.1998.tb10780.x.

57. Seidel, C., & Bicker, G., (1996). The developmental expression of serotonin immunoreactivity in the brain of the pupal honeybee. *Tissue Cell, 28*, 663–672. doi: 10.1016/s0040-8166(96)80070-x.
58. Seidel, C., & Bicker, G., (2002). Developmental expression of nitric oxide/cyclic GMP signaling pathways in the brain of the embryonic grasshopper. *Brain Res. Dev. Brain Res., 138*, 71–79. doi: 10.1016/s0165-3806(02) 00466-2.
59. Siegl, T., Schachtner, J., Holstein, G. R., & Homberg, U., (2009). NO/cGMP signaling: L-citrulline and cGMP immunostaining in the central complex of the desert locust *Schistocerca gregaria. Cell Tissue Res., 337*, 327–340. doi: 10. 1007/s00441-009-0820-z.
60. Sitaraman, D., Zars, M., Laferriere, H., Chen, Y. C., Sable-Smith, A., Kitamoto, T., et al., (2008). Serotonin is necessary for place memory in *Drosophila. Proc. Natl. Acad. Sci. U S A, 105*, 5579–5584. doi: 10.1073/pnas.0710168105.
61. Strausfeld, N. J., (1976). *Atlas of an Insect Brain.* Berlin, Heidelberg, New York, NY: Springer.
62. Strausfeld, N. J., (1999). A brain region in insects that supervises walking. *Prog. Brain Res., 123*, 273–284. doi: 10.1016/s0079-6123(08)62863-0.
63. Strausfeld, N. J., (2012). *Arthropod Brains.* Cambridge, MA: Harvard Univ. Press.
64. Strauss, R., (2002). The central complex and the genetic dissection of locomotor behavior. *Curr. Opin. Neurobiol., 12*, 633–638. doi: 10.1016/s09594388(02)00385-9.
65. Sykes, P. A., & Condron, B. G., (2005). Development and sensitivity to serotonin of *Drosophila* serotonergic varicosities in the central nervous system. *Dev. Biol., 286*, 207–216. doi: 10.1016/j.ydbio.2005.07.025.
66. Taghert, P. H., & Goodman, C. S., (1984). Cell determination and differentiation of identified serotonin-immunoreactive neurons in the grasshopper embryo. *J. Neurosci., 4*, 989–1000.
67. Taghert, P. H., & Truman, J. W., (1982a). The distribution and molecular characteristics of the tanning hormone, bursicon, in the tobacco hornmoth, *Manduca sexta. J. Exp. Biol., 98*, 373–383.
68. Taghert, P. H., & Truman, J. W., (1982b). Identification of the bursicon containing neurons in abdominal ganglia of the tobacco hornworm moth, *Manduca sexta. J. Exp. Biol., 98*, 385–401.

69. Tettamanti, M., Armstrong, J. D., Endo, K., Yang, M. Y., Furukubo-Tokunaga, K., Kaiser, K., et al., (1997). Early development of the *Drosophila* mushroom bodies, brain centers for associative learning and memory. *Dev. Genes Evol., 207*, 242–252. doi: 10.1007/s004270050112.

70. Thompson, K. S. J., & Siegler, M. V. S., (1991). Anatomy and physiology of spiking local and intersegmental interneurons in the median neuroblast lineage of the grasshopper. *J. Comp. Neurol., 305*, 659–675. doi: 10.1002/cne.903050409.

71. Thompson, K. S. J., Rayne, R. C., Gibbon, C. R., May, S. T., Patel, M., Coast, G. M., et al., (1995). Cellular colocalization of diuretic peptides in locusts: A potent control mechanism. *Peptides, 16*, 95–104. doi: 10.1016/0196-9781(94) 00158-3.

72. Tomer, R., Denes, A. S., Tessmar-Raible, K., & Arendt, D., (2010). Profiling by image registration reveals common origin of annelid mushroom bodies and vertebrate pallium. *Cell, 142*, 800–809. doi: 10.1016/j.cell.2010.07.043.

73. Triphan, T., Poeck, B., Neuser, K., & Strauss, R., (2010). Visual targeting of motor actions in climbing *Drosophila. Curr. Biol., 20*, 663–668. doi: 10.1016/j.cub.2010. 02.055.

74. Truman, J. W., De Vente, J., & Ball, E. E., (1996a). Nitric oxide-sensitive guanylate cyclase activity is associated with the maturational phase of neuronal development in insects. *Development, 122*, 3949–3958.

75. Truman, J. W., Ewer, J., & Ball, E. E., (1996b). Dynamics of cyclic GMP levels in identified neurones during ecdysis behavior in the locust *Locusta migratoria. J. Exp. Biol., 199*, 749–758.

76. Turlejski, K., (1996). Evolutionary ancient roles of serotonin: Long-lasting regulation of activity and development. *Acta Neurobiol. Expt. (Wars), 56*, 619–636.

77. Urbach, R., & Technau, G. M., (2003). Early steps in building the insect brain: Neuroblast formation and segmental patterning in the developing brain of different insect species. *Arthropod Struct. Dev., 32*, 103–123. doi: 10.1016/s14678039(03)00042-2.

78. Vitalis, T., Cases, O., Passemard, S., Callebert, J., & Parnavelas, J. G., (2007). Embryonic depletion of serotonin affects cortical development. *Eur. J. Neurosci., 26*, 331–344. doi: 10.1111/j.1460-9568.2007.05661.x.

79. Vitzthum, H., & Homberg, U., (1998). Immunocytochemical demonstration of locustatachykinin-related peptides in the central complex of the locust brain. *J. Comp. Neurol., 390*, 455–469. doi: 10.1002/(SICI)10969861(199 80126)390:4<455::AID-CNE1>3.0.CO;2-#.

80. Vitzthum, H., Homberg, U., & Agricola, H., (1996). Distribution of Dip-allatostatin I-like immunoreactivity in the brain of the locust *Schistocerca gregaria* with detailed analysis of immunostaining in the central complex. *J. Comp. Neurol., 369*, 419–437. doi: 10.1002/(SICI)10969861(19960603)369:3<419::AID-CNE7>3.0.CO;2-8.
81. Wegerhoff, R., & Breidbach, O., (1992). Structure and development of the larval central complex in a holometabolous insect, the beetle *Tenebrio molitor. Cell Tissue Res., 268*, 341–358. doi: 10.1007/bf00318803.
82. Wegerhoff, R., Breidbach, O., & Lobemeier, M., (1996). Development of locustatachykinin immunopositive neurons in the central complex of the beetle *Tenebrio molitor. J. Comp. Neurol., 375*, 157–166. doi: 10.1002/(SICI)10969861(19961104)375:1<157::AID-CNE10>3.0.CO;2-S.
83. Wehner, R., (1989). Neurobiology of polarization vision. *Trends Neurosci., 12*, 353–359. doi: 10.1016/0166-2236(89)90043-x.
84. Weir, P. T., & Dickinson, M. H., (2012). Flying *Drosophila* orient to sky polarization. *Curr. Biol., 22*, 21–27. doi: 10.1016/j.cub.2011. 11.026.
85. Westbrook, A. L., & Bollenbacher, W. E., (1990). The development of identified neurosecretory cells in the tobacco hornworm, *Manduca sexta*. *Dev. Biol.* 140, 291–299. doi: 10.1016/0012-1606(90)90079-x.
86. Williams, J. L. D., (1975). Anatomical studies of the insect central nervous system: A ground-plan of the midbrain and an introduction to the central complex in the locust, *Schistocerca gregaria* (Orthoptera). *J. Zool. Lond., 176*, 67–86.doi: 10. 1111/j.1469-7998.1975.tb03188.x.
87. Williams, J. L. D., Güntner, M., & Boyan, G. S., (2005). Building the central complex of the grasshopper *Schistocerca gregaria*: Temporal topology organizes the neuroarchitecture of the w, x, y, z tracts. *Arthr. Struct. Dev., 34*, 97–110. doi: 10.1016/j.asd.2004.11.001.
88. Williams, J. L., & Boyan, G. S., (2008). Building the central complex of the grasshopper *Schistocerca gregaria*: Axons pioneering the w, x, y, z tracts project onto the primary commissural fascicle of the brain. *Arthr. Struct. Dev., 37*, 129–140. doi: 10.1016/j.asd.2007.05.005.
89. Winther, A. M., Acebes, A., & Ferrus, A., (2006). Tachykinin-related peptides modulate odor perception and locomotor activity in Drosophila. *Mol. Cell. Neurosci., 31*, 399–406. doi: 10.1016/j.mcn.2005.10.010.
90. Wolff, T., Iyer, N. A., & Rubin, G. M., (2015). Neuroarchitecture and neuroanatomy of the *Drosophila* central complex: A GAL4-based dissection of protocerebral bridge neurons and circuits. *J. Comp. Neurol., 523*, 997–1037. doi: 10.1002/cne.23705.
91. Wong-Riley, M. T., & Liu, Q., (2005). Neurochemical development of

brain stem nuclei involved in the control of respiration. *Respir. Physiol. Neurobiol., 149*, 83–98. doi: 10.1016/j.resp.2005.01.011.

92. Young, J. M., & Armstrong, J. D., (2010a). Structure of the adult central complex in *Drosophila*: Organization of distinct neuronal subsets. *J. Comp. Neurol., 518*, 1500–1524. doi: 10.1002/cne.22284.
93. Young, J. M., & Armstrong, J. D., (2010b). Building the central complex in *Drosophila*: The generation and development of distinct subsets. *J. Comp. Neurol., 518*, 1525–1541. doi: 10.1002/cne.
94. Yuan, Q., Lin, F., Zheng, X., & Sehgal, A., (2005). Serotonin modulates circadian entrainment in *Drosophila. Neuron, 47*, 115–127. doi: 10.1016/j.neuron.2005.break05.027.
95. Zacharias, D., Williams, J. L. D., Meier, T., & Reichert, H., (1993). Neurogenesis in the insect brain: Cellular identification and molecular characterization of brain neuroblasts in the grasshopper embryo. *Development, 118*, 941–955.
96. Zars, T., (2009). Spatial orientation in *Drosophila. J. Neurogenet., 23*, 104–110. doi: 10.1080/01677060802441364.
97. Zitnan, D., Ross, L. S., Zitnanova, I., Hermesman, J. L., Gill, S. S., & Adams, M., (1999). Steroid induction of peptide hormone gene leads to orchestration of a defined behavioral sequence. *Neuron, 23*, 523–535.

CHAPTER 4

Neurochemistry and the Fundamentals of Neurotransmitters

Contents

4.1. INTRODUCTION

Neurotransmitters (NTs) are chemical messengers that allow neurons to interact with one another. Considering the brain's complexity, it must not be surprising that there are over 100 NTs identified (Purves et al., 2008). NTs are made in nerve cells and injected into the synapse, wherein they bind to particular receptors on the postsynaptic cell, occasionally requiring dietary precursors (e.g., tyrosine). This simplified picture hides the numerous intriguing ways in which neurons connect, as detailed in this chapter. We will focus on the many types of NTs, as well as their production, transport, storage, and release, as well as their activity at deactivation and receptors. Drugs' effects on NT function would be highlighted as well.

4.2. THE NEURON AND THE SYNAPSE

Figure 4.1 depicts a typical neuron. Neurons have a cell body that houses the nucleus as well as an axon and the distinctive dendrites. Dendrites receive signals from other cells and store them on their shafts and spines, whereas the axon sends data to other cells. While each nerve cell already has 1 axon, it can have numerous branches, as well as the nerve endpoints at every branch's end may make synapses with certain other neurons (Ando et al., 2017). The discharge of neurotransmitters (NTs) from the nerve terminals of the axon into the synapse, the region between the postsynaptic and presynaptic cells, allows nerve cells to communicate with one another. The postsynaptic cell's receptors connect to NTs released into the synapse. Synapses may occur in between the numerous distinct places on the postsynaptic cell and axon of the presynaptic cell, such as the shafts and/or spines of dendrites (axodendritic synapses), the axons (axodendritic synapses), and the cell body (axosomatic synapses), as illustrated in Figure 4.1. Synapses may be inhibitory or excitatory. Upon an excitatory synapse, the NT produced by the presynaptic cell stimulates the postsynaptic cell, enabling it to become depolarized and create an action potential along the axon by changing the electrical potential of the cell membrane. The NT emitted by the presynaptic cell prevents the postsynaptic cell from activating at inhibitory synapses. Inhibitory synapses are located on the axon terminals and cell bodies, whereas excitatory synapses are located on dendrites (Zador et al., 1993; Gallagher et al., 1995).

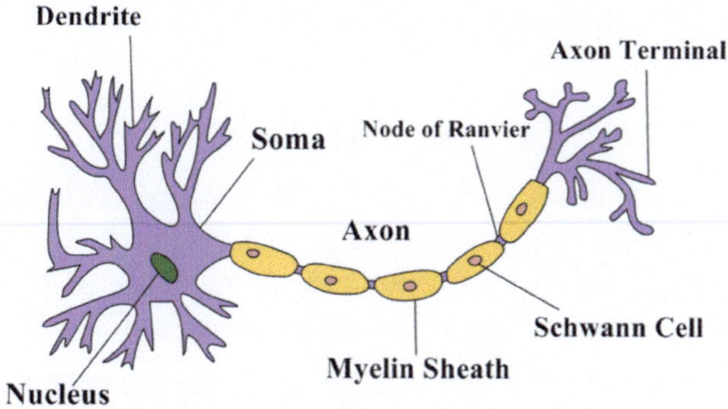

Figure 4.1. *A typical neuron's structure.*

Note: Dendrites are complicated branches on neurons that collect messages from other neurons (A, B, and C) through synapses. Synapses (D–axodendritic synapses) can be found on dendritic spines or dendritic shafts. The cell body (soma; F–axosomatic synapses) and Axons (E–axoaxonic synapses) both have synapses. Connections are formed with others through the neurons nerve terminal (H) and a single axon (G), which emits a neurotransmitter into the synaptic space. The position of the action potentials is shown by large black arrows.

Source: https://www.vedantu.com/question-answer/draw-a-labeled-diagram-of-the-neuron-and-class-11-biology-cbse-5f7da6c8605c2300d9c992b7.

4.3. WHAT IS A NEUROTRANSMITTER (NT)?

A NT is a chemical messenger that transports information from one neuron to another cell, which might or could not be another neuron. Many neurons, for instance, regulate other neurons, although a few also regulate endocrine glands and muscles (Eckstein et al., 2020). Autoreceptors, also known as presynaptic receptors, allow neurons to react to NTs via autocrine signaling. To be classified as a NT, a neurochemical must fulfill specific requirements that separate "real" NTs from substances with neuronal function. Smokers, cocaine, and heroin addicts, for example, have excessive quantities of nicotine and drugs in their brains. In addition, psychiatric patients rely on antidepressant and anti-anxiety drugs' capacity to work in the brain. To achieve their enjoyable or therapeutic results, all of such chemicals disrupt normal neurotransmission, although they are not NTs (Gerra et al., 1997; Stanford, 2001). A chemical compound must fulfill many requirements to be classified as a NT. Such characteristics apply to both classical NTs like neuropeptide and acetylcholine (ACh) NTs like opioids or substance P

(Squire et al., 2008). Nestler et al. (2009); Iversen et al. (2010) offer a more extensive overview of the brain's NT chemicals (2009).

4.4. CATEGORIES OF NEUROTRANSMITTERS (NTS)

A vast variety of compounds may now be classified as NTs using the criteria described in the following points. The primary NTs are divided into four types, as shown in Table 4.1: monoamines, amino acids, ACh, and peptides (Purves et al., 2008). Criteria for determining that a neuroregulatory molecule is a NT are as follows:

- The neurotransmitter is generated in a neuron, the precursors and synthesis enzymes should be present, for instance, the enzymatic conversion of the precursor tyrosine to dopamine (DA). As a result, DA neurons should have both tyrosine hydroxylase and tyrosine.
- The distribution of neurotransmitters in the brain is uneven. They are only present in certain places and are linked to certain neural pathways. Opioid peptides, for instance, are frequently found as a neurotransmitter in brain areas linked with pain.
- The material should be present in the terminals of presynaptic neurons, generally in the form of vesicles.
- The substance-binding receptors must be located near the presynaptic neuron terminals. When released neurotransmitters bind to such receptors, alterations in postsynaptic neurons must occur. The drug must, for instance, cause changes in the permeability of the postsynaptic membrane, resulting in inhibitory or excitatory potentials in the postsynaptic cell.
- Stimulation of a neuron path or neuron must cause in physiologically sufficient amounts of the chemical being released from the nerve terminals of neurons.
- The neurotransmitter's effects must be terminated by certain methods. Enzymatic breakdown or reuptakes into the presynaptic terminal are examples of such processes.
- Interventions at postsynaptic locations, such as the use of particular receptor antagonist medications, must prevent the impacts of nerve stimulation, whilst agonist drugs must replicate the neurotransmitter's functions.
- Direct injection of the chemical into incorrect synapses in the

brain must elicit effects that are equivalent to those induced by activating neuronal pathways.

4.4.1. The Amino Acid Neurotransmitters (NTs)

In the central nervous system (CNS) of mammalian, amino acid NTs are the most plentiful and are found in neurons throughout the spinal cord and brain. The excitatory NT glutamate, for instance, is linked to practically each physiological brain route and is engaged in pathological processes like epilepsy, brain damage by addiction and stroke. Ion channels in the membrane of their postsynaptic target cells are amino acid NT receptors (Walz, 1985; Altschuler and Shore, 2010).

Table 4.1. Categories of Neurotransmitters

Category	Neurotransmitter
Cholinergic neurotransmitter	Acetylcholine (ACh)
Amino acid transmitters Inhibitory Excitatory	Gamma-aminobutyric acid (GABA), glycine Glutamic acid Aspartic acid
Peptide neurotransmitters	Somatostatin Substance P Enkephalins, β-endorphin (opioids) Cholecystokinin (CCK) Neurotensin
Monoamine neurotransmitters Dopaminergic Catecholamines Indoleamine Adrenergic Other	Norepinephrine (NE; sometimes noradrenaline) Dopamine (DA) Serotonin (5-HT) Epinephrine (or adrenaline) Histamine

Putative neurotransmitters	Prostaglandins ATP (adenosine triphosphate), adenosine
Unusual neurotransmitters	Endogenous cannabis: Endocannabinoids Gases: nitric oxide and hydrogen sulfide Zinc; D-serine

Such channels are opened as a result of the amino acids binding to them, resulting in fast inhibitory or excitatory responses on the postsynaptic cells. -aminobutyric acid (GABA), for instance, functions as an inhibitory NT by opening Cl^- ion channels through the action of GABA receptors in the brain (Figure 4.2). It has the effect of reducing electrophysiological activity in postsynaptic neurons.

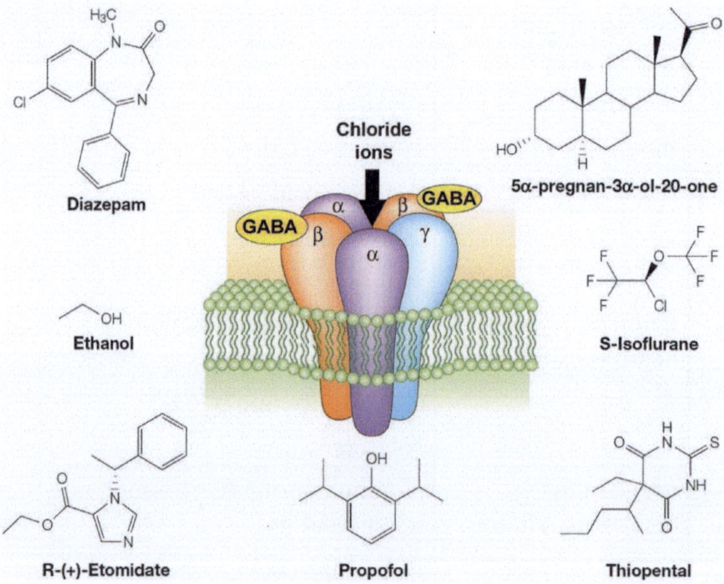

Figure 4.2. *The inhibitory GABA receptor's structure.*

Note: This Cl^- ion channel is found in neuron cell membranes and is composed of five subunits (α, β, and γ).

Source: https://www.nature.com/articles/nrn1703.

Because of this inhibition, the impacts of anesthetics (for example, thiopental; propofol), tranquilizers (such as Diazepam), alcohol, and barbiturates may all be observed, and they are all mediated by GABA receptors (Belelli and Lambert, 2005; Nestler et al., 2009). It is worth noting that glycine is another important amino acid inhibitory transmitter. Glycine functions through a separate ion channel, which similarly regulates the entrance of Cl^- ions into the postsynaptic neuron. Its principal locations of activity are in the spinal cord, which is where it originated (Bustamante et al., 2002).

GABA binds to the receptor, allowing Cl^- ions to enter the cell through the central channel. Medications including isoflurane, thiopental, and propofol as well as anti-anxiety drugs like diazepam and alcohol, have binding sites in the receptor complex. Such medications work by boosting chloride entrance into the cell, which enhances the actions of GABA. In a variety of physiological and pathophysiological circumstances, neurosteroids like 5α-pregnan-3α-ol-20-one may affect behavior and mood (Belelli and Lambert, 2005).

Glycine has a role in excitatory neurotransmission as well (shown in Figure 4.3). In the spinal cord and brain, the excitatory amino acids aspartic acid (aspartate) and glutamic acid (glutamate) serve as NTs (Nestler et al., 2009; Purves et al., 2008). They are ubiquitously located in the brain, and it is believed that glutamate is used as a NT by around half of all brain synapses. Unlike GABA, aspartate, and glutamate receptors, certain of which are also ion channels that cause excitement by opening postsynaptic channels for Na^+ ion and Ca^{+2} ions to enter (Figure 4.3).

Due to the widespread distribution of such excitatory NTs throughout the brain, this is not unexpected that they are linked to a wide range of physiological activities. Induction of long-lasting variations in synaptic function (also known as synaptic plasticity) is among the most significant and well-studied effects of glutamate action. Synaptic plasticity serves as the foundation for the neuronal alterations that underlie memory and learning.

4.4.2. Acetylcholine (ACh)

Acetylcholine (ACh) is released by the neurons of the cholinergic paths when they are stimulated. Along with working broadly across the CNS and brain, ACh is the NT that is employed at the neuromuscular junction, in the parasympathetic branch of the autonomic nervous system (ANS), and the autonomic ganglia, among other places. ACh is a NT that has a role in memory and attention functions, and it can be implicated in age-related disorders like Alzheimer's disease (AD) (Nicholls et al., 2001). ACh is also involved in a variety of motivated behaviors, such as sexual behavior, aggressiveness, and the control of drinking and thirst behavior. Another receptor for ACh is the nicotinic receptor, which functions as an ion channel. However, ACh may also act through the muscarinic receptor (Jensen et al., 2005; Xue et al., 2021).

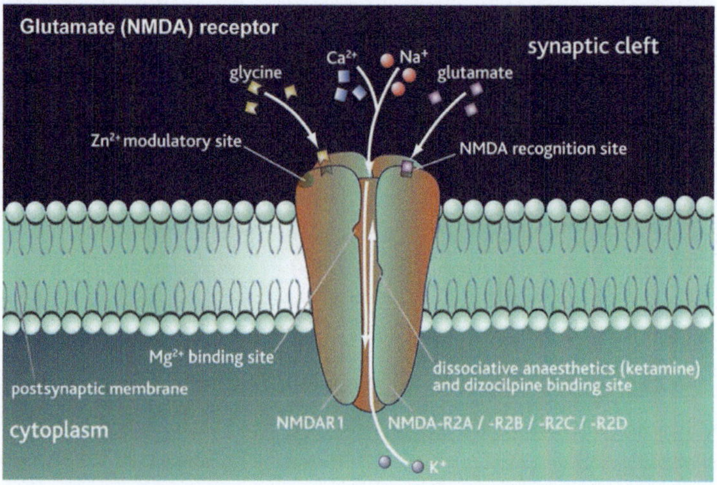

Figure 4.3. *Structure of an excitatory glutamate receptor.*

Note: Excitatory glutamate receptors are ion channels that allow Na^{+} ion and Ca^{2+} ions to enter the target neuron, causing it to depolarize and become excitable. The glutamate receptor subtype NMDA (N-methyl-D-aspartate) is used for illustration. For instance, the NMDA receptor is composed of four subunits that are organized around a central pore, with NMDAR1 in conjunction with multiple NMDAR2 subunits being the most commonly encountered (NMDA-R2A, -R2B, -R2C or -R2D). In addition to glutamate, the receptor complex is influenced by several additional substances, which are listed below. As a result, glycine is a key co-agonist, as it is required for glutamate to open the channel pore and allow Na^{+} ion and Ca^{2+} ions to enter the cell to function properly. Also required is the bonding of zinc ions to the channel, while magnesium ions operate to block the channel till depolarization happens. Additionally, anesthetics and hallucinogenic drugs like dizocilpine and ketamine, among others, act on the receptor to produce their effects.

Source: https://www.cambridge.org/core/books/abs/an-introduction-to-neuroendocrinology/neurotransmitters/F1AD7D06C5B3BEE1F-4092028A0727FDC.

4.4.3. The Monoamine Neurotransmitters (NTs)

Monoamine NTs are found in considerably lower concentrations in the brain as compared to amino acid NTs, although they are essential as drug objectives in the treatment of anxiety, depression, Parkinson's disease, and drug addiction (Nestler et al., 2009). Whenever these NTs engage to their receptors, they trigger a series of chemical changes in the cell's cytoplasm including 2nd messenger systems (Nestler et al., 2009).

4.4.4. Peptide Neurotransmitters (NTs)

Neuropeptides, which operate as NTs in the CNS were first found in the digestive system. Numerous central neuronal circuits include extremely low quantities of vasoactive intestinal polypeptide (VIP), cholecystokinin (CCK), bombesin, and gastrin. Neuropeptides such as neurotensin, substance P, enkephalins, and somatostatin are generated in neurons and fulfill the majority of the NT action requirements. The way neuropeptides and traditional NTs like DA are generated by neurons is a significant distinction. Large propeptides are digested to produce smaller, active peptides, which are then used to make neuropeptides. Enkephalins, or endogenous opioids, are generated initially as big proenkephalins in the brain and spinal cords neuron cell bodies. Enkephalins, the end products, are tiny molecules with just 5 amino acids that behave like NTs, especially in the control of pain perception. Other neuropeptides operate as neuromodulators, modifying how a cell reacts to some other NT like DA (Madhavaiah and Verma, 2005; Shiao et al., 2017).

4.4.5. Non-Classical Neurotransmitters (NTs)

Thus far, all of the compounds described have met the majority of the requirements for being classified as NTs. Other neurochemical substances, like the gas nitric oxide (NO) and cannabis-like compounds (endocannabinoids), may serve as NTs while not meeting many of the characteristics stated in Table 4.1. Nitric oxide is produced in neurons when the NT glutamate activates the enzyme nitric oxide synthase (NOS; Figure 4.4; Sullivan, 2003), which converts the amino acid arginine to nitric oxide. Although just 1% of brain neurons produce this enzyme, nitric oxide is thought to be present in practically every neuron (Squire et al., 2008), emphasizing the critical role of nitric oxide in the function of the brain. Because nitric oxide is a gas, it is liberated from neurons by diffusion rather than being stored in secreted/vesicles. Furthermore, while nitric oxide travels backwards across the synapse (retrograde action) to modulate glutamate production in presynaptic neurons, it also diffuses out from the neuron terminal to affect glial cells and neurons far away. Nitric oxide also varies from other NTs since it does not bind to particular receptors; instead, it diffuses into other cells and influences signaling systems like cyclic guanosine monophosphate (cGMP) (Baranano et al., 2001; Vincent, 2010).

As a result, a minimum of three of the conditions listed in Table 4.1 are unmet. Nonetheless, nitric oxide is now widely regarded as a NT. Other gases, including CO and hydrogen sulfide, have also been shown to behave as NTs (Olson, 2011; Deng et al., 2014).

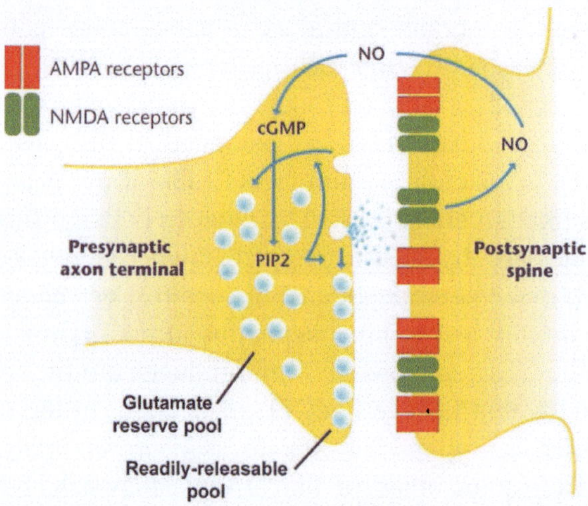

Figure 4.4. *Nitric oxide (NO) is an extraordinary neurotransmitter.*

Note: In response to the emission of glutamate from pre-synaptic axon terminals, the NMDA receptors on postsynaptic nerve terminals are activated, causing the release and production of nitric oxide. Another glutamate receptor subtype, known as AMPA receptors, is also activated in this manner. Nitric oxide works as a retrograde messenger, boosting the synthesis of cGMP and PIP2 at the presynaptic terminal as a result of its action. The rise in PIP2 speed up the rate during which vesicles are collected and prepared for re-use in the readily-releasable pool, which is beneficial in many situations. As a result, nitric oxide increases glutamate neurotransmission.

Abbreviations: NO: nitric oxide; AMPA: α-amino-3-hydroxy-5-methyl-4-isoxazole propionic acid; PIP2: phosphatidylinositol diphosphate; cGMP: cyclic guanosine monophosphate. Reproduced by authorization (Sullivan, 2003).

Source: https://www.ncbi.nlm.nih.gov/books/NBK10795/.

Although the euphoric qualities of cannabis were recognized for thousands of years, the quest for brain cannabis receptors (CB1 receptors) has only lately proven effective. The identification of endogenous brain chemicals that mirrored the activity of tetrahydrocannabinol was made possible by the revelation that the brain has receptors specialized for a plant molecule (THC). Endocannabinoids were named after these chemical messengers, and anandamide (AEA) is the most common member of this class. AEA, as nitric oxide, is created enzymatically only when required (in other words, this is not stored) and is easily diffusible across cell membranes, therefore it is not secreted. It appears to have a key role as a feedback mechanism for presynaptic neurons. Thus, it is a retrograde messenger that inhibits NT release by acting on presynaptic CB receptors (Figures 4.5 and 4.6; Guzmán, 2003; Christie and Vaughn, 2001).

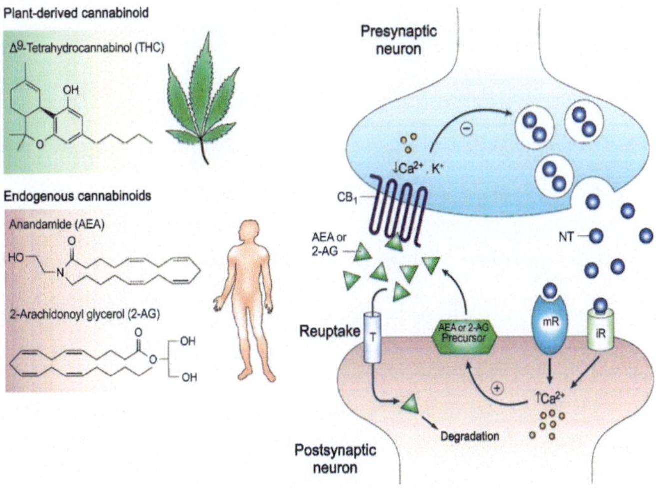

Figure 4.5. *Neurotransmitters derived from endogenous cannabis-like substances.*

Note: Endocannabinoids, like 2-arachidonoyl glycerol (2-AG) or anandamide (AEA), are synthesized in the postsynaptic cell membrane via precursors by neurotransmitter (NT) activation of metabotropic membrane receptors (mR) or ion channels (iR). Anandamide (or 2-AG) is released into the synaptic cleft and interacts with presynaptic cannabinoid receptors that are G-protein coupled (CB1). As a consequence, neurotransmitter release is inhibited. Excessive AEA is reabsorbed into the postsynaptic cell through a transporter (T), accompanied by the enzymatic breakdown (Guzmán, 2003).

Source: https://pubmed.ncbi.nlm.nih.gov/14570037/.

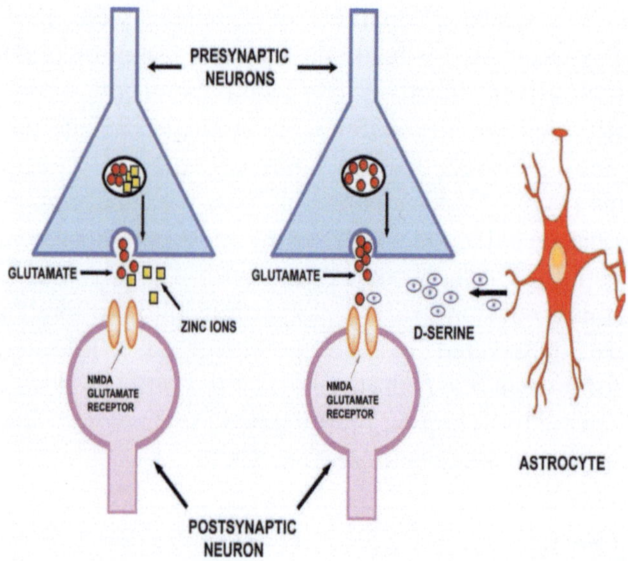

Figure 4.6. *D-serine and zinc ions are two uncommon neurotransmitters.*

Note: Zinc ions engage to glutamate receptors of the N-methyl-D-aspartate (NMDA) subcategory because they are co-released and so co-localized with glutamate. D-serine, on either side, is produced by astrocytes and attaches to the NMDA receptor to increase glutamate activity, most likely at the glycine site.

Source: https://www.cambridge.org/core/books/an-introduction-to-neuroendocrinology/C988ABF8E5B25B4BCAB77B6BD94E5B3B.

The various impacts of cannabis on mood, memory, hunger, coordination, pain sensation, and sleep are due to the widespread distribution of CB receptors throughout the brain. CB receptors can be involved in obesity while marijuana stimulates appetite and Rimonabant, CB receptor blocker decreases the weight of the body and eating (Christensen et al., 2007; Ward and Raffa, 2011).

4.5. NEUROTRANSMITTER (NT) BIOSYNTHESIS AND STORAGE

The manufacturing of monoamine NTs and amino acids differs significantly from the manufacturing of neuropeptides. On the ribosomes of the endoplasmic reticulum in the cell body of the neuron, enzymes for the production of amino acids and amines are generated and stored in synaptic vesicles. Such synaptic vesicles also contain amino acid substrates like tryptophan (for serotonin) and tyrosine (for DA) (Palmada and Centelles, 1998). While they travel down the axon to the nerve terminals, the NTs' production is finished within synaptic vesicles. The precursors (pre-propeptides) of neuropeptide NTs are also synthesized in the cell body and kept in vesicles containing enzymes (convertases) that process the precursor peptides by cutting them into relatively small active molecules like the vesicles are transmitted to the nerve terminals (Figure 4.7). Synaptic vesicles are stored in a storage pool within the nerve terminal before being liberated into the synapse (Bartolome et al., 1982; Opladen et al., 2016):

- Synaptic vesicles serve numerous important functions in nerve cells: They transmit biosynthetic enzymes and neurotransmitter propeptides/precursors from the cell body to the nerve terminal along the axon;
- The manufacturing of neurotransmitters is frequently finalized inside the vesicles;
- The vesicles store neurotransmitters until they have been released;
- Vesicles defend neurotransmitters from discontinuation;
- Neurotransmitters are released from nerve terminals via exocytosis, which occurs when a vesicle merges with a depolarized cell membrane in the existence of Ca^{+2} ions, and
- Vesicles may exert feedback control on neurotransmitter production, in which enzymes like tyrosine hydroxylase have been prompted to generate further norepinephrine when vesicles are depleted of the neurotransmitter. Once norepinephrine levels in the vesicles are returned to normal, the vesicle acts as a -ve feedback impact on tyrosine hydroxylase, shutting off future norepinephrine synthesis.

4.5.1. Storage and Synthesis of Certain Neurotransmitters (NTs)

4.5.1.1. Amino Acid Neurotransmitters (NTs)

In glial and neurons cells, aspartate, and glutamate are made using glucose and other precursors. The enzyme glutamic acid decarboxylase (GAD) in nerve terminals produces GABA from its precursor, glutamic acid. The production of amino acid NTs is covered extensively below (Nestler et al., 2009).

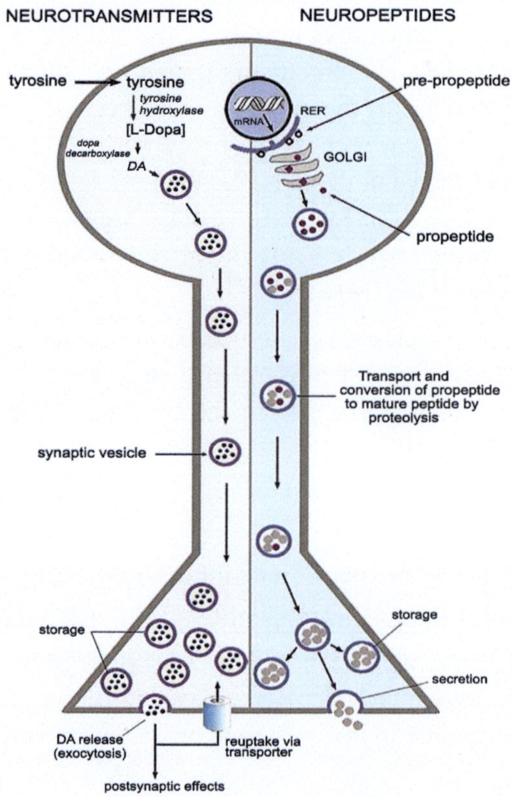

Figure 4.7. *Biosynthesis of neurotransmitters vs. neuropeptides.*

Note: Neuropeptide biosynthesis is based on gene translation and transcription in the cell body, with ultimate structural modification within the vesicles. Classical neurotransmitters, on the other hand, rely on the re-absorption of dietary precursors like tyrosine. DA is synthesized in the cytoplasm from the precursor tyrosine and afterwards stored in vesicles before being transported to the nerve terminal. Norepinephrine is bio-synthesized from DA within vesicles in certain neurons. Another significant distinction among classical neurotransmitters and neuropeptides is that the latter cannot be utilized after discharge, as there is no re-absorption mechanism. There is an effective re-absorption system for classical neurotransmitters which provides a constant supply of fully packed vesicles prepared for emission.

Abbreviations: mRNA: messenger ribonucleic acid; DA: dopamine; RER: rough endoplasmic reticulum.

Source: https://www.amazon.com/Introduction-Neuroendocrinology-Michael-Wilkinson/dp/052101476X.

4.5.1.2. Acetylcholine (ACh)

Acetylcholine is produced using choline, an amino acid absorbed from the circulation and acetyl coenzyme through the activity of the enzyme choline acetyltransferase (CAT) (Nestler et al., 2009; Iversen et al., 2009). CAT is made in the cell body and carried down the axon to the nerve terminals, wherever it is used to synthesize ACh (Figure 4.8) (Racagni et al., 1975; Lee and Park, 2014).

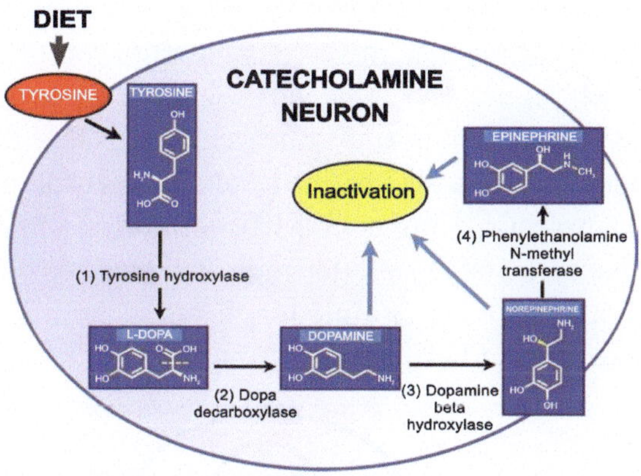

Figure 4.8. *Catecholamines biosynthesis.*

Note: The structural variation that happens when tyrosine is converted to three distinct catecholamines is depicted in this diagram. DA neurons use the enzyme tyrosine hydroxylase (1) to convert tyrosine to L-Dopa, which is subsequently transformed to DA through dopa decarboxylase. (2) Norepinephrine neurons follow the identical procedures as DA neurons, with the exception that the enzyme DA beta-hydroxylase. (3) Inserts another hydroxyl group to produce norepinephrine. Ultimately, phenylethanolamine N-methyltransferase helps epinephrine neurons transform norepinephrine to epinephrine through adding a methyl group. (4) Catecholamines are inactivated by enzymes in the same way.

Source: https://www.nhbs.com/an-introduction-to-neuroendocrinology-book.

4.5.2. Regulation of Neurotransmitter (NT) Synthesis and Storage

Synthesizing, storing, releasing, and deactivating NTs is a constant process. Moreover, because the NT should not be depleted at the neuron terminal, there are numerous methods for maintaining an appropriate quantity of releasable NTs. The quickest and most reliable method to refill the storage vesicles in a norepinephrine neuron, for instance, is to recover norepinephrine that is already released into the synapse (Kruk and Pycock, 1991; Halbach and Dermietzel, 2006). The re-absorption process in the neuron terminal works as a vacuum cleaner, sucking up any free norepinephrine from the synaptic space and returning it to the terminal, and it is absorbed by the vesicles. The re-absorption is not being enough to replenish the vesicles in situations when the neuron is constantly releasing and firing huge quantities of norepinephrine. In this situation, whether the vacant vesicles signal tyrosine hydroxylase to create additional norepinephrine or the nucleus is signaled to boost tyrosine hydroxylase mRNA expression in severe conditions (Iversen et al., 2009; Figure 4.9). A negative feedback loop inhibits tyrosine hydroxylase (end product inhibition) once enough NT is deposited in the vesicles. NT production may also be influenced by neurological input from other cells (for example, Ca^{+2} ions or cAMP), hormones, and the accessibility of amino acids like tyrosine. The latter indicates that our diet may have an impact on certain aspects of brain function in some circumstances (Richard et al., 2009; Fernstrom and Fernstrom, 2007).

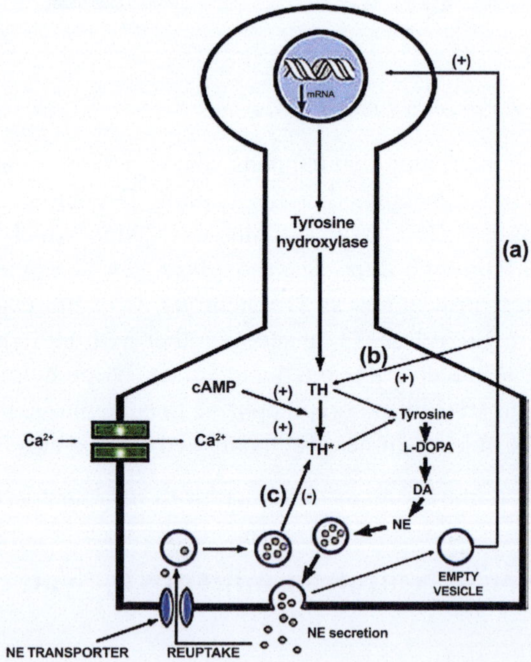

Figure 4.9. *Maintaining sufficient catecholamine levels in nerve terminals.*

Note: This figure utilizes norepinephrine (NE) as an example, although the explanation applies to dopamine as well. Following the release of significant quantities of norepinephrine by excessive neuronal activity, appropriate levels of norepinephrine are restored first by absorption through the norepinephrine transporter, and then by enhanced norepinephrine synthesis in the neuron terminal. When vesicles lose their norepinephrine content, the empty vesicle transmits two signals: (a) to the nucleus, which activates the tyrosine hydroxylase (TH) gene and the production of additional tyrosine hydroxylase; and (b) to the terminal, which activates existing tyrosine hydroxylase (such as TH to TH*). This cycle is likewise true in reverse; (c) when vesicles are replenished, tyrosine hydroxylase activity is turned off.

Abbreviations: cAMP: cyclic adenosine monophosphate; DA: dopamine.

Source: https://www.sciencedirect.com/topics/medicine-and-dentistry/catecholamine.

4.6. RELEASE OF NEUROTRANSMITTERS (NT)

Neurons are defined as both a presynaptic cell and a post-synaptic; that is, when a neuron is triggered postsynaptically by NTs from another neuron, the neuron discharges its NT presynaptically into the next synapse; this is known as the postsynaptic-release hypothesis. Such NTs excite their postsynaptic receptors at excitatory synapses, resulting in the opening of ion channels and the depolarization of the neuronal membrane, in which sodium (Na+) flows into the cell and potassium (K+) exits from the cell. When the cell depolarizes, it produces a detectable shift in its electrical activity, which is referred to as an action potential (Santos et al., 1996). The action potential travels down the nerve cell membrane until it arrives at the nerve terminal of the axon (Figure 4.10).

A change in the permeability of the cell membrane happens when depolarization of the nerve terminal occurs, resulting in the release of NTs under the influence of calcium ions. The ratio of the release of NTs to those released in intracellular fluid is reversed when depolarization of the nerve terminal takes place (Figure 4.11) (Miller et al., 1982; Rhee et al., 2005).

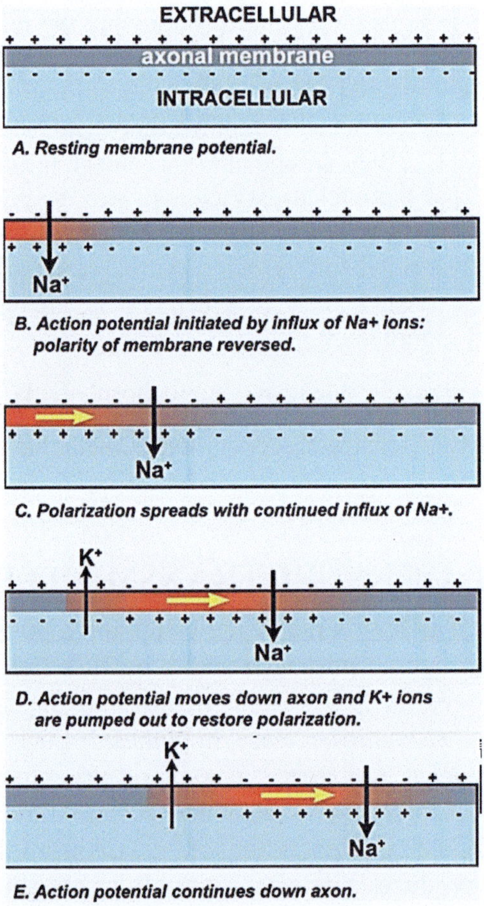

Figure 4.10. *The production of an action potential.*

Note: A localized input of sodium (Na^+) ions accompanied by an outflow of potassium (K^+) ions through channels regulated by voltage variation around the axonal membrane corresponds with the transmission of a nerve impulse (action potential) by the axon. Through segregating fluids with different concentrations of sodium ion and potassium ion (A); the cell membrane creates a resting membrane potential. Sodium ions are 10 times more abundant in extracellular fluid as compared to potassium ions. In most cases, the electrical event that sends a nerve impulse down the axon starts in the cell body. The action potential starts with depolarization of the axonal membrane, or a decrease in the -ve potential. A few of the sodium ion channels are opened as a result of the voltage change. The inflow of sodium ions increases till the membrane's inner surface changes from –ve to +ve charge (B). The sodium ion channel shuts and the potassium channel open as a result of this voltage reversal (C). The potassium ion outflow promptly recovers the axons -ve potential, and the action potential propagates along the axon (D and E). A 2nd impulse may flow after a brief refractory phase wherein the potassium ion is pumped into the cell and the sodium ion is pumped out of the cell.

Source: https://www.researchgate.net/publication/290497385_An_introduction_to_neuroendocrinology_second_edition.

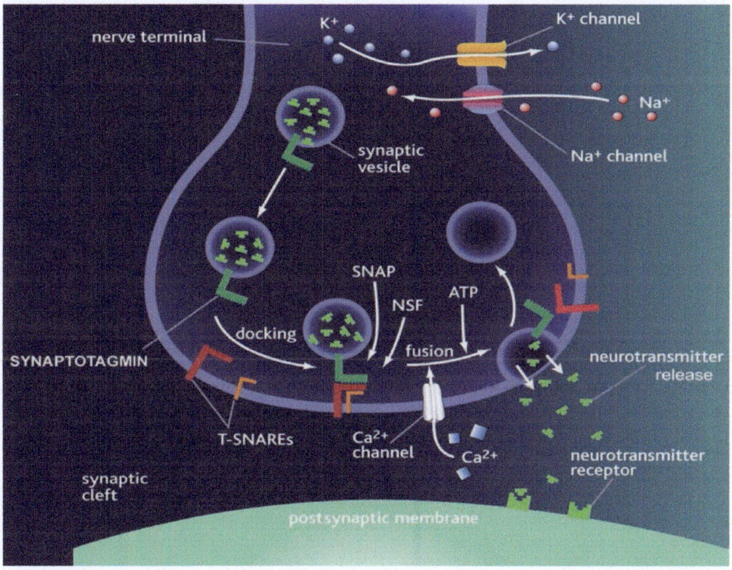

Figure 4.11. *The method through which vesicles release neurotransmitters.*

Note: Following the fusing of the synaptic vesicle membrane with the neuronal membrane, neurotransmitters are liberated from the nerve terminal. This fusion occurs due to the presence of certain proteins such as NSF, synaptotagmin, SNARE, and SNAP. NSF is a fusion protein, whereas SNAP is an attachment protein for NSF; T-SNARE is a SNAP receptor, whereas synaptotagmin is a SNAP vesicle attachment protein. Notably, action potential transmission (through sodium ion and potassium ions) results in Ca^{2+} influx across the presynaptic membrane, and membrane fusion occurs as a result of SNARES, synaptotagmin, and other factors.

Abbreviations: ATP: adenosine triphosphate; NSF: N-ethylmaleimide sensitive factor; SNAP: soluble SNF attachment proteins; SNARE: SNAP receptor protein.

Source: https://www.researchgate.net/figure/MP-regulates-neurotransmitter-release-MP-dephosphorylates-syntaxin-at-Ser14-and-SNAP-25_fig2_326778580.

Synaptic vesicle membrane elements are regenerated following Neurotransmitters (NTs) release through an endocytic procedure.

4.7. RECEPTORS FOR NEUROTRANSMITTERS (NTS)

After being injected into the synapse, the NT affects postsynaptic neuronal processes by attaching to a particular receptor protein or ion channel on the postsynaptic cell's surface membrane (Figure 4.12).

Because the majority of NTs are too polar (water-soluble) to reach their target cells (for example, they cannot cross the phospholipid cell membrane), they excite receptors placed on the cell membrane's periphery. Although, there are certain outliers, like endocannabinoids and nitric oxide, which move readily across cell membranes. Membrane receptors are classified into three categories: ionotropic receptors that open ion channels and thus are principally activated by amino acid NTs (Figure 4.13) (Okuyama et al., 1997; Rose et al., 2008).

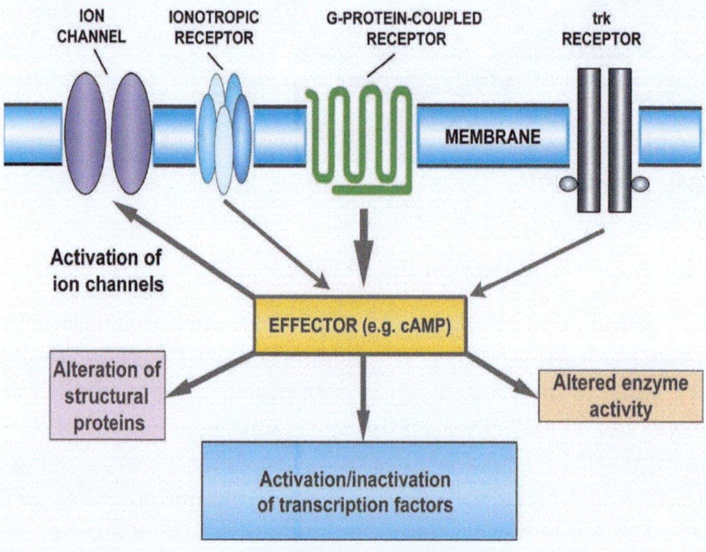

Figure 4.12. *Generalized mechanisms for cellular activation.*

Note: The activation of neurons is regulated by a variety of ion channels and receptors. The neurotransmitter norepinephrine, for instance, engages to a metabotropic receptor protein (also known as a G-protein-coupled receptor) that stimulates a variety of cellular activities through intracellular effectors like cAMP. Trk receptors bind hormones like leptin, whereas ionotropic receptors react to neurotransmitters like glutamate.

Abbreviations: cAMP: cyclic adenosine monophosphate; Trk: tyrosine receptor kinase.

Source: https://www.cambridge.org/core/books/an-introduction-to-neuroendocrinology/1C450D6854B43294B9E89F201B8BC6BC.

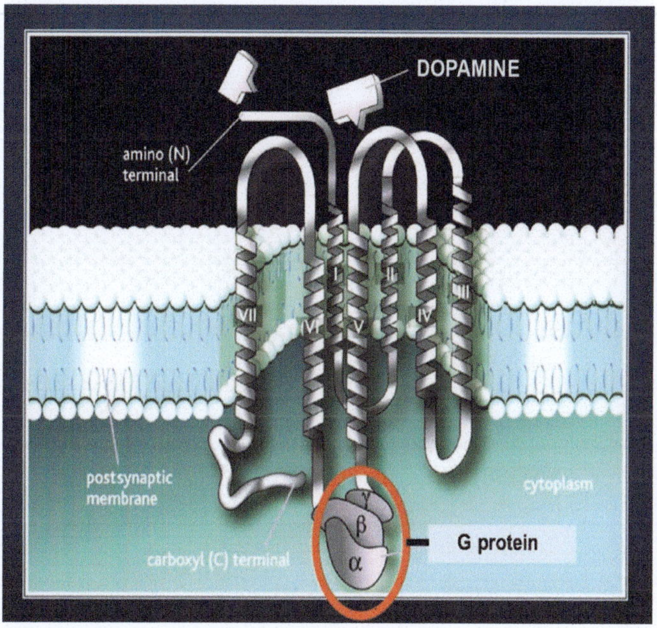

Figure 4.13. *G-protein-coupled receptor structure.*

Note: G-protein-coupled receptors are found on the surface of several neurotransmitters, such as DA. Such membrane receptors have seven transmembrane spanning α-helices that cross the membrane. G proteins are activated by DA binding to the extracellular region of the receptor, which results in the activation of intracellular secondary messenger signaling pathways. A G protein is a complicated composed of three subunits denoted by α, β, and γ. The downstream impact would be either inhibitory or stimulatory based upon the type of G protein which is connected to the receptor–dopamine D1, D5 receptors are connected to stimulatory G proteins (Gs), although dopamine D2, D3, and D4 are connected to inhibitory G proteins (Gi).

Source: https://www.researchgate.net/figure/Schematic-drawing-of-a-G-protein-coupled-receptor-structure_fig1_234048401.

In addition to GABA (Figure 4.2) and glutamate receptors (Figure 4.3), there are metabotropic receptors that activate intracellular 2nd messenger systems like cAMP to monitor and control metabolic variation inside the postsynaptic cell, and which are stimulated by peptide and catecholamine NTs. Such receptors are also referred to as G-protein-coupled receptors in some circles (Figure 4.13). Tyrosine receptor kinase (trk receptor) is the name given to the 3rd kind of receptor, which is responsible for the action of neurotrophins (for example, NGF), growth factors, and adipokines (for example, leptin). The ligand, like leptin, binds to these single-stranded proteins, which project through the cell membrane and link together (dimerize) when they connect. Table 4.2 shows the subtypes of NT receptors.

Table 4.2. Neurotransmitter Receptor Subtypes

SL. No.	Receptor Sub-Types
1.	**GABAergic** $GABA_A$, $GABA_B$
2.	**Glutamatergic** NMDA, AMPA kainate, $mGluR_1$, $mGluR_2$, $mGluR_3$, $mGluR_4$, $mGluR_5$, $mGluR_6$
3.	**Glycinergic** Glycine
4.	**Cholinergic** Muscarinic: M_1, M_2, M_3, M_4, M_5 Nicotinic: muscle, neuronal (α-bungarotoxin-insensitive)
5.	**Dopaminergic** D1, D2, D3, D4, D5
6.	**Adrenergic** α1A, α1B, α1C, α1D, α2A, α2B, α2C, α2D, β1, β2, β3
7.	**Serotonergic** $5\text{-}HT_{1A}$, $5\text{-}HT_{1B}$, $5\text{-}HT_{1D}$, $5\text{-}HT_{1E}$, $5\text{-}HT_{1F}$, $5\text{-}HT_{2A}$, $5\text{-}HT_{2B}$, $5\text{-}HT_{2C}$, $5\text{-}HT_3$, $5\text{-}HT_4$, $5\text{-}HT_{5A}$, $5\text{-}HT_{5B}$, $5\text{-}HT_6$, $5\text{-}HT_7$
8.	**Histaminergic** H1, H2, H3, H4
9.	**Opioid** μ1, μ2, μ3, δ1, δ2, κ1, κ2, κ3
10.	**Endocannabinoid** CB1 (brain), CB2 (peripheral tissues)

Abbreviations: 5-HT: 5-hydroxytryptamine; AMPA: α-amino-3-hydroxy-5-methyl-4-isoxazole propionic acid; GABA: γ-aminobutyric acid; mGluR: metabotropic glutamate receptor.

There are several kinds of receptors for every NT, which are referred to as receptor subtypes. The studies of these receptors, known as receptor neuropharmacology, are a vast field. Table 4.2 gives a quick rundown of the number of receptor subtypes for the NTs we have just covered. Subtypes of receptors enable cells to respond to the same NT in various ways. Whenever norepinephrine attaches to 1-adrenergic receptors, for instance, blood vessel smooth muscle cells contract, whereas when NE attaches to 2-adrenergic receptors, they relax (Purves et al., 2008). DA receptor subtypes can elicit quite diverse physiological responses; for instance, DA D1 receptors increase cAMP synthesis, while D2 receptors decrease cAMP synthesis. For the development of novel medications, understanding receptor subtypes is essential. Novel anti-migraine medicines, for instance, have been discovered that bind to the 5-HT1B/1D receptor subtype selectively. Many receptor subtypes have already been meticulously mapped in terms of their proteins (Kuhar et al., 1986) and the cellular locations of their genes (Mansour et al., 1995; Van den Pol et al., 1994).

4.8. INACTIVATION OF NEUROTRANSMITTERS (NTS)

Efficient neurotransmission necessitates not just NT production from the presynaptic terminal, and also efficient and faster deactivation. It is significant since a NT's continual stimulation of postsynaptic receptors may be harmful to neurons. Uncontrolled activation of glutamate receptors, for instance, may result in neuronal death and convulsions. To prevent this, NTs are swiftly deactivated through a variety of mechanisms (Suzuki et al., 1994). Physical removal of the NT by re-absorption into the presynaptic terminal through transporters is the most effective method. NTs may also be destroyed enzymatically (Nestler et al., 2009; Iversen et al., 2009). Figure 4.14 illustrates these methods for norepinephrine. The most effective route to eliminate norepinephrine from the synapse is by re-absorption through the repackaging and transporter in vesicles. This phase also recycles any NTs that are no longer in use, reducing the requirement for more manufacturing. Reuptake of neuropeptides into presynaptic terminals does not tend to happen (Di Guilmi et al., 2011). Because of the significance of transporters in synaptic transmission, it was discovered that illegal drugs like amphetamine and cocaine work by binding to DA transporters and preventing DA reabsorption and deactivation. To put it another way, the stimulant impact (the "rush") of such drugs is generated by DA lingering in the synapse for longer than typical. The mode of action of various antidepressant drugs, like Prozac, is another instance. This drug inhibits serotonin reabsorption and raises serotonin levels in the synapse. There are several such instances of the effects of certain drugs that disrupt transporters (McCabe, 1986).

4.9. NEUROTRANSMITTER (NT) PATHWAYS

In the brain of human beings, NTs are found in distinct neuronal pathways. Figure 4.15 depicts DA, NE, serotonin, ACh, glutamate, and GABA circuits (Tiwari et al., 2104).

DA pathways are concentrated into three main systems (Figure 4.15(A)): (1) The substantia nigra projects to the caudate nucleus/putamen (also known as the striatum); (2) the ventral tegmentum projects to the cerebral cortex and different limbic structures like the hippocampus, septum, nucleus accumbens, and amygdala; and (3) the hypothalamic short tuberoinfundibular system (Iversen et al., 2009; Nestler et al., 2009).

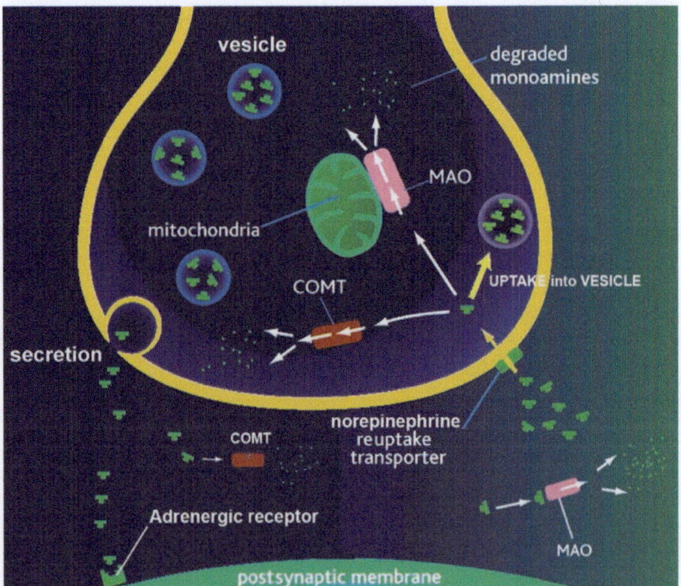

Figure 4.14. *Inactivation mechanisms for catecholamines.*

Note: Catecholamines are inactivated primarily by specialized transporters that remove them from the synapse and allow them to be packed in secretory vesicles. This phase also recycles any neurotransmitters that are no longer in use, reducing the requirement for more synthesis. Catecholamines are also fastly destroyed by two enzymes: catechol o-methyl transferase (COMT) and monoamine oxidase (MAO). It is worth noting that the enzymes can be found both within and outside the neuron terminal (in the synapse). Although all neurotransmitters possess degrading enzymes, it is not like all neurotransmitters possess particular transporters (Ong and Holbrook, 2014; Meadows et al., 2019).

Abbreviations: COMT: catechol o-methyl transferase; MAO: monoamine oxidase.

Source: https://www.ncbi.nlm.nih.gov/books/NBK217810/.

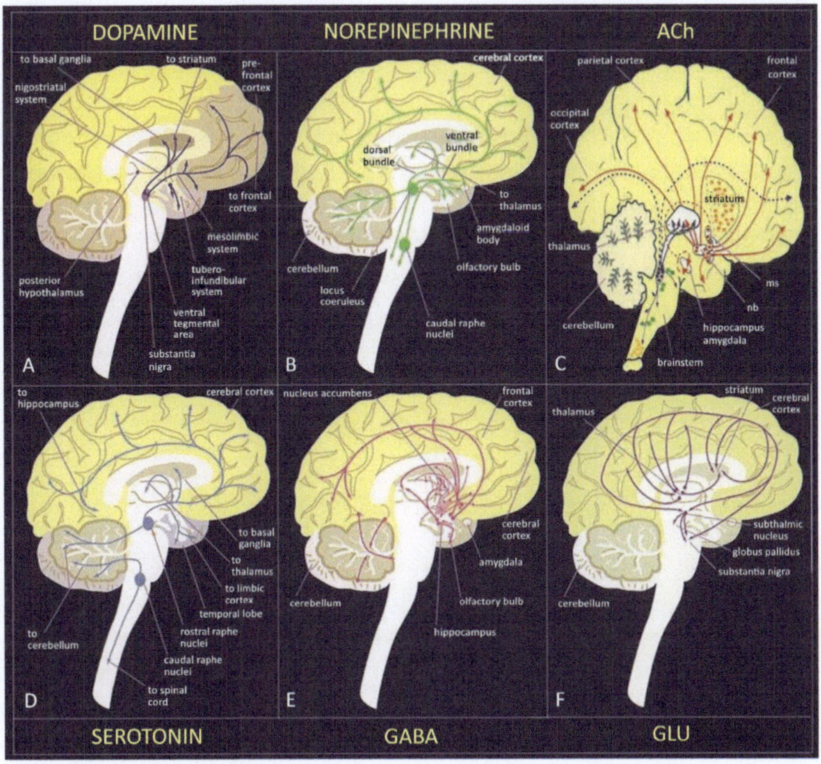

Figure 4.15. *Neuronal routes for neurotransmitters in the brain.*

Note: The location of projecting axons and neuronal cell bodies for six neurotransmitters are depicted in this figure: dopamine (A); norepinephrine (B); acetylcholine (ACh; C); serotonin (D); GABA (E); and glutamate (Glu; F). DA has conveyed three main passageways: (i) an extension from the substantia nigra to the caudate nucleus-putamen (neostriatum); (ii) a passageway connecting the ventral tegmentum to the mesolimbic forebrain; and (iii) a short passageway called the tuberoinfundibular system, which is involved in the neuronal control of the hypothalamic-pituitary system. Numerous parts of the brain are fed by norepinephrine (noradrenergic) systems, which are located in the caudal Raphé nucleus and the locus ceruleus. The locus ceruleus' ascending nerves terminate in the hypothalamus, thalamus, frontal cortex, and limbic system. Additionally, norepinephrine is delivered from the locus ceruleus to the cerebellum via the locus ceruleus. The caudal Raphé's nerves descend to the midbrain and ascend to the amygdala. Acetylcholine is distributed throughout the brain via two main pathways: basal-forebrain cholinergic neurons (red, such as medial septal nucleus (ms) and the nucleus basalis (nb)) and pedunculopontine–lateral dorsal tegmental neurons (blue). Additionally, striatal interneurons (orange), cranial nerve nuclei (green circles), and spinal-cord preganglionic and motor neurons are cholinergic neurons (yellow). The rostral and caudal Raphé nuclei are the primary locations of serotonin neurons. Axons ascend from the rostral nuclei to the cerebral cortex, limbic areas, and most especially the basal ganglia. Caudal Raphé serotonin nuclei generate descending axons, a few of which end in the medulla and others in the spinal cord. GABA is the primary inhibitory neurotransmitter in the central nervous system, and its inhibitory effects are shown at all levels, including the hippocampus, hypothalamus, cerebral cortex, and cerebellar cortex. Along with the well-established GABA routes, GABA interneurons are numerous in the brain, accounting for 50% of inhibitory synapses. Glutamate is the most common excitatory neurotransmitter, and this diagram illustrates the primary connections between the cortex and thalamus. Furthermore, glutamate receptors are abundant all across the brain, notably in the hypothalamus, showing that numerous brain locations are innervated by glutamate.

Source: https://www.pinterest.com/pin/11681280259484678/.

The DA route between the substantia nigra and the caudate nucleus is crucial for movement integration. The loss of DA cells in the substantia nigra induces stiffness and tremor in Parkinson's disease patients. By allowing the manufacturing of extra DA, injections of L-Dopa, a DA precursor that may cross the blood-brain barrier (shown in Figure 4.8), alleviate most of these signs. The DA system is also a component of a complicated reward route and a target for drugs of misuse like amphetamine. It has also been linked to the cause of attention deficit hyperactivity disorder (ADHD) (ADHD).

The hypothalamic tuberoinfundibular DA pathway is responsible for a major portion of the control of prolactin release from the anterior pituitary.

In addition to the locus ceruleus, norepinephrine neurons are present in a variety of other brain regions (Figure 4.15(B)), and they are one of the most significant places of the cell bodies in the locus ceruleus. It is composed of two noradrenergic networks: a dorsal medial forebrain bundle and a ventral medial forebrain bundle. These channels innervate a variety of tissues, including the olfactory bulb, cortex, amygdala, hypothalamus, and thalamus, among others.

The cerebellum and spinal cord are also innervated by additional descending pathways (Iversen et al., 2009; Nestler et al., 2009). Norepinephrine in the brain has been implicated in the regulation of the sleep-wake cycle, attention, eating behavior, and as a constituent of normal fear and stress reactions, among other functions. In addition, there is strong evidence that it has a role in depressive diseases. Noradrenergic neurons influence motivated and emotional behavior through their projections to the hypothalamus and median eminence, and they also regulate the secretion of hypothalamic inhibiting/releasing hormones, which in turn regulate the secretion of pituitary hormones like luteinizing hormone (LH), growth hormone, and prolactin.

Figure 4.15(C) depicts the neuronal passageways that produce ACh (Perry et al., 1999). There are two main projections exist: (1) Cholinergic neurons (CN) in the basal forebrain have been found in the nucleus basalis (nb) and the medial septal nucleus (ms), and these neurons innervate extensive areas of the cerebral cortex; and (2) CNs in the brainstem may innervate the thalamus as well as motor neurons in the spinal cord. The 3rd type of neuron is found in the striatum and is known as interneurons. As a site of cholinergic regulation, the hypothalamus is also a site of NT categorization, which reveals that ACh is responsible for stimulating the release of hormones such as ACTH, thyroid-stimulating hormone, and growth hormone, as well as the neurons that secrete vasopressin and oxytocin. Other studies have shown that ACh has a direct impact on GnRH neurons, causing them to secrete the neuropeptide (Krsmanovic et al., 1998).

It is virtually solely from cell bodies in the Raphé nuclei that serotonin pathways (the serotoninergic system; Figure 4.15(D)) stretch, with axons ascending to most parts of the brain. These areas include the hypothalamus, the hippocampus, cerebral cortex, and amygdala. Axons originating in the caudal aspect of the Raphé nuclei descend to the spinal cord and cerebellum (Nestler et al., 2009). This NT is associated with a broad range of brain processes, as evidenced by the extensive dispersion of serotonin axons and the presence of many serotonin receptors throughout the brain. Sleep and wakefulness, anxiety, sadness, the impact of psychedelic substances, and migraine are among the issues addressed (Iversen et al., 2009). Similarly, serotonin neurons have been linked in the regulation of anterior pituitary hormone secretion, but they can also be involved in the regulation of posterior pituitary hormone (for example, vasopressin) secretion (Jorgensen, 2007).

NTs made from amino acids are important signaling molecules in the brain. They are divided into two categories: (1) excitatory NTs like glutamate and aspartate, and (2) inhibitory NTs like GABA and glycine. In the spinal cord and brain, GABA is the most common inhibitory NT, and several GABA neurons occur as local circuit inhibitory interneurons or neurons with extremely short axons which emit GABA to inhibit neighboring neurons. Moreover, certain GABAergic neurons have longer axons that branch out to other parts of the brain (Figure 4.15(E)). The thalamus, hippocampus, and cortex are all innervated by cell bodies in the nucleus accumbens/substantia nigra. The secretion of pituitary and hypothalamic hormones is also regulated by GABAergic neurons. Glutamate is the brain's primary excitatory NT, and Figure 4.15(F) depicts various glutamate passageways. Moreover, the illustration only shows a portion of the glutamate neuron projections (Maccarrone and Wenger, 2005; Pagotto et al., 2006; Rettori et al., 2010). Because of that, the brain contains enormous amounts of glutamate that are not associated with NT function; that is, glutamate is a precursor in the manufacture of GABA, and glutamate also plays a vital role in neuronal metabolism in addition to its activity as a NT (Iversen et al., 2009). As a result, the placement of glutamate receptors is more indicative of glutamate's substantial significance as a NT in brain function. A method termed *in situ* hybridization was used to map the distribution of glutamate receptor subtypes in neurons. For instance, glutamate receptors of the NMDA subtype (shown in Figure 4.3) have been found in the olfactory bulb, thalamus, cortex, and hippocampus (Watanabe et al., 1993), as well as in the hypothalamus (Eyigor et al., 2001). The latter findings are by glutamate's significant role in pituitary hormone secretion that is controlled by the hypothalamus (Van den Pol et al., 1990; Brann, 1995). Glutamate, for instance, is an important aspect of the hypothalamic regulation of fertility via stimulatory impacts on GnRH production, as well as the neurochemical control of adrenal and thyroid hormone secretion through TRH and CRH neurons (Iremonger et al., 2010; Wittman et al., 2005; Levy and Tasker, 2012). Glutamate has been linked to the regulation of neuronal plasticity, particularly as a foundation for learning and memory (Watkins and Jane, 2006), and both GABA and glutamate have been linked to a variety of mental diseases. GABA dysfunction, for instance, has been associated with epilepsy, Huntington's disease, and sleep disturbances, while glutamate has been connected to epilepsy, stroke, and neurotoxicity (de Kloet and Woods, 2009; Iversen et al., 2009). Gaseous NTs like nitric oxide are not kept in vesicles and instead diffuse away from respective production sites. As a result, pinpointing individual brain circuits is challenging. The enzyme NOS, which is responsible for the manufacture of nitric oxide, may be utilized to identify neurons that produce nitric oxide. Immunohistochemistry was used to map NOS neurons across the rat brain, finding NO-specific neurons in the olfactory bulbs, cerebral cortex, amygdala, hippocampus, cerebellum, and substantia nigra (Rodrigo et al., 1994; Vincent, 2010; Hill and Tasker, 2012). The hypothalamus has the largest concentration of NOS in the brain, and nitric oxide is involved in pituitary hormone secretion (Prevot et al., 2000; Brann et al., 1997).

4.10. DRUGS INFLUENCING NEUROTRANSMITTERS (NTS), AND THEIR RECEPTORS, IN THE NERVOUS SYSTEM

There are a variety of medicines that affect receptor activation and NT release. As we know the nervous system's normal functioning has come through the usage of drugs that interact with certain NT receptors (Fowler et al., 1984). Treatments for some nervous system problems of mood, motor function, emotionality, neuroendocrine secretion, and drug misuse have come from the information gained by utilizing drugs as instruments to investigate brain function (Nestler et al., 2009; Iversen et al., 2009).

Drugs may affect NT action in at least three ways: (1) they may act as neuromodulators, influencing the synthesis, release, storage, and reabsorption of the NT by the presynaptic cell; (2) they may change the activity of deactivating enzymes in the synapse, trying to influence the synaptic concentrations of the NT; or (3) they may mimic (or block) the NT's impact at the synapse. Some of these pharmacological activities are mentioned as examples. Drugs manufactured to treat certain illnesses may have various side effects, some of which include the neuroendocrine system since they impact all parts of the brain using a particular NT system (Sasse, 2007; Pritchett et al., 2010).

4.10.1. Drugs Altering Neurotransmitter (NT) Synthesis, Storage, Release, and Reuptake

Extensive animal experiments have proven that the levels of NTs may be altered through the use of different neuropharmacological approaches. Examples include reserpine, which significantly lowers norepinephrine levels in the brain by blocking its absorption into vesicles, therefore enabling breakdown to take place (reabsorption into vesicles depicted in Figure 4.14). Drugs are also available that lower NT levels by inhibiting the manufacturing of the NTs (Komsta and Wicha-Komsta, 2018). When NT levels need to be raised, as in Parkinson's disease, patients have been treated with L-Dopa, a precursor to DA that allows the striatum to synthesize extra DA. This is effective in increasing NT levels. There is also evidence that some nutrients in our diet, like tryptophan and tyrosine, may have an effect on NT levels in the brain. By inhibiting the typical deactivation mechanisms, the amounts of NTs in the synaptic cleft may be raised. For instance, medicines like clorgyline inhibit the degrading enzyme monoamine oxidase (MAO; shown in Figure 4.16), while entacapone inhibits COMT (catechol o-methyl transferase), both of which result in an increase in synaptic norepinephrine levels in the brain. Other drugs (such as Prozac) enhance both norepinephrine and serotonin levels by inhibiting their reabsorption through their particular transporters (Figure 4.16). Figure 4.17 depicts the impact of a number of these drugs on postsynaptic cellular responses (Sundvik, 2012).

A large number of drugs that change NT levels or activity are listed in Table 4.3, certain of which are utilized in therapeutic settings and may have an impact on elements of endocrine function (Visser, 2012).

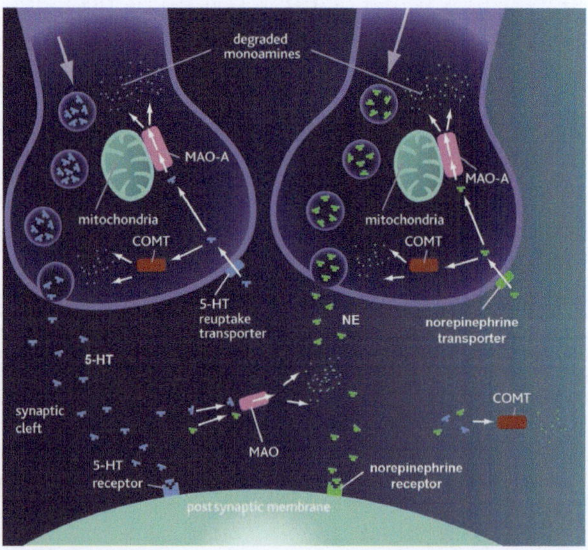

Figure 4.16. *Blockade of norepinephrine and serotonin reabsorption.*

Note: This diagram depicts how neurotransmitter levels, like norepinephrine (NE) and serotonin (5-HT), are controlled and how they may be altered by therapeutic medicines like those utilized to treat depression. Depression, for instance, is related to decreased levels of the neurotransmitters norepinephrine and serotonin in the brain. Norepinephrine and serotonin reabsorption inhibitors, like Prozac, are hypothesized to work by preventing the reabsorption and subsequent degradation of such neurotransmitters in the synaptic cleft, therefore restoring their levels in the cleft. The buildup of catecholamines in the synaptic cleft, as a result of this reuptake inhibition, theoretically leads to the concentration of catecholamines returning within the normal range. It appears from the diagram that inhibition of monoamine oxidase (MAO) or catechol o-methyl transferase (COMT)-induced degradation will likewise result in an increase in the amounts of norepinephrine and serotonin at synaptic sites.

Source: https://www.pharmacology2000.com/nursing_pharmacology/Central/Antidepressant2N/Depression-7.htm.

4.10.2. Drugs That Act at Postsynaptic Receptors

Another way in which synaptic transmission can be altered is by using drugs that bind to postsynaptic receptors and act as NT agonists or antagonists. Agonist drugs can mimic or even magnify the effect of the NT (Vizi, and Kiss, 1998; Groc et al., 2007).

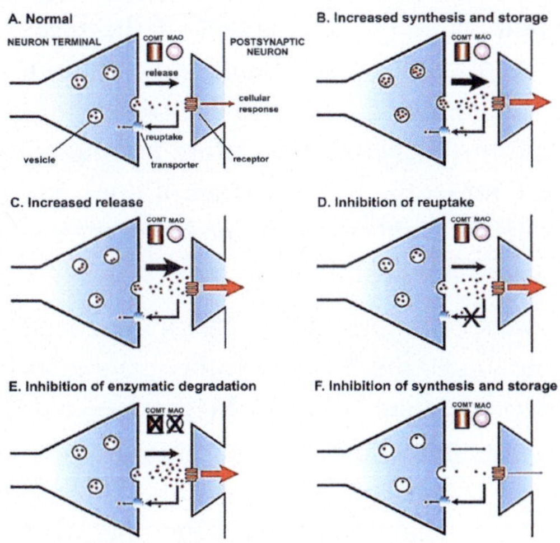

Figure 4.17. *The methods by which medications affect the number of synaptic catecholamine neurotransmitters accessible in the synaptic synapse, as well as the resultant postsynaptic cellular response.*

Note: (A) Storage, reuptake, release, and enzymatic metabolism of norepinephrine (by MAO enzymes) all cause a postsynaptic cellular response (red arrow) through a postsynaptic G-protein-coupled receptor in the usual condition Storage, reuptake, release, and enzymatic metabolism of norepinephrine (by MAO enzymes) all cause a postsynaptic cellular response (red arrow) through a postsynaptic G-protein-coupled receptor in the usual condition. (B) Increased storage and synthesis, such as after tyrosine or L-Dopa therapy, can boost NE or DA release and postsynaptic cellular response (larger arrows). (C) An augmented postsynaptic cellular response is induced by increased norepinephrine release, such as after greater nerve stimulation or therapy with amphetamine. (D) Antidepressants like imipramine, for instance, impede reuptake, which raises synaptic norepinephrine levels, resulting in an increased postsynaptic cellular response. (E) Inhibitors of MAO- or COMT-induced catecholamine metabolism, like clorgyline, raise the postsynaptic cellular response and synaptic levels of the neurotransmitter. (F) The inhibition of norepinephrine storage (by reserpine) or inhibition of synthesis minimizes synaptic levels of neurotransmitters and thus inhibits the postsynaptic cellular response.

Abbreviations: COMT: catechol o-methyl transferase; MAO: monoamine oxidase.

Source: https://www.cambridge.org/core/books/abs/an-introduction-to-neuroendocrinology/neurotransmitters/F1AD7D06C5B3BEE1F-4092028A0727FDC.

An agonist would raise the excitement of the postsynaptic cell at an excitatory synapse. An agonist would enhance inhibition if the synapse is inhibitory. A NT antagonist, on either side, works by blocking the NT's receptors. The antagonist's chemical structure permits it to engage with the receptor, but it does not excite the cell when it does so (Jaen et al., 1988). Because the antagonist prevents the genuine NT at the synapse from binding to the receptor, it would not be able to do so.

A NT agonist works similarly to the NT: it attaches to the postsynaptic cell and either stimulates or inhibits it. Synthetic agonists have the potential to be even more effective than NTs. LSD (lysergic acid diethylamide), for instance, is a serotonin agonist that produces psychedelic effects by binding to 5-HT2A serotonin receptors (Iversen et al., 2009). Other psychedelic drugs, such as PCP (phencyclidine) and ketamine, bind as antagonists to glutamate receptors of the NMDA (N-methyl-D-aspartate) subtype.

Table 4.3. Drugs Change the Number of Neurotransmitters Available in the Synapse, and They are Utilized to Treat Several Neurological Conditions (Iversen et al., 2009; Nestler et al., 2009)

Transmitter	Drug	Action
Dopamine	Tyrosine and L-Dopa (Parkinson's) Clorgyline Entacapone (Parkin-son's) Deprenyl (Parkinson's) Cocaine Amphetamines Reserpine Ritalin (ADHD) α-methyltyrosine	(+) increase dopamine synthesis (+) inhibits MAOA (+) inhibits MAOB (+) blocks reuptake; in-creases release (+) inhibits COMT (+) blocks reuptake (+) blocks reuptake vesicle uptake (–) inhibits dopamine synthesis (–) causes depletion by inhibiting
GABA	Tiagabine (epilepsy) Valproate (epilepsy; bipolar disorder) Gabapentin (epilepsy; pain) Vigabatrin (epilepsy) Tetanus toxin Benzodiazepine (anxi-ety) Allylglycine	(+) blocks reuptake (+) increases GABA func-tion (+) enhances GABA func-tion (+) blocks GABA trans-aminase (–) inhibits release (+) enhances GABA func-tion (–) inhibits synthesis
Acetylcho-line	Donepezil (Alzheimer's) Choline Botulinum toxin Black Widow spider venom Pyridines Chantix (smoking ces-sation) Scopolamine (motion sickness)	(+) acetylcholinesterase inhibitor (+) stimulates synthesis and release (–) inhibits release (+) promotes release (–) block synthesis (+) nicotinic agonist/an-tagonist (–) blocks muscarinic receptor

<table>
<tr><td>Serotonin</td><td>Tryptophan
Iproniazid
Iproniazid
Prozac (antidepressant)
Ecstasy (recreational drug)
p-chlorophenyl alanine
Reserpine
LSD</td><td>(+) increases synthesis
(+) inhibits MAO
(+) blocks reuptake
(+) blocks reuptake
(+) blocks reuptake and is a neurotoxin
(–) inhibits synthesis
(–) depletion by blocking reuptake into vesicles
(–) depletion by blocking reuptake into vesicles</td></tr>
<tr><td>Norepinephrine</td><td>Cocaine
Imipramine (tricyclic antidepressant)
Effexor (antidepressant)
Amphetamines
Entacapone
Clorgyline; deprenyl
α-methyltyrosine
Reserpine</td><td>(+) blocks reuptake
(+) blocks reuptake
(+) blocks reuptake
(+) MAO inhibition
(+) mixed NE/5-HT uptake inhibition
(–) inhibits synthesis
(+) COMT inhibition
(–) depletes stores by inhibiting vesicle reuptake</td></tr>
</table>

4.10.3. Drugs – The Neuroendocrine System and Behavior

In addition to affecting NT release and function, drugs that modify emotional states, arousal levels, and motivated behavior can also have an impact on hormone secretion. Since low catecholamine levels are related to depression, and excessive catecholamine levels are associated with schizophrenia, drugs that change catecholaminergic activity also have an impact on these clinical symptoms (Kostoff and Lau, 2013; Lee et al., 2018). Changes in catecholamine levels have been shown to affect both sexual and violent motivation, as well as the functioning of the neuroendocrine system. In general, drugs that affect NT levels in the brain can also have neuroendocrine adverse effects like decreased hunger, decreased sexual desire, and decreased prolactin release (Papakostas et al., 2006).

4.11. CAN NUTRIENTS MODIFY NEUROTRANSMITTER (NT) LEVELS AND BEHAVIOR?

The availability of precursor amino acids from the diet is required for the biosynthesis of several NTs, like serotonin and norepinephrine. Richard Wurtman and his colleagues started this study (Wurtman, 1982; Milner and

Wurtman, 1986), and Fernstrom and coworkers have continued it (Fernstrom et al., 2007; Choi et al., 2009; Fernstrom, 2012). They believe that amino acids in the diet affect the rate at which the NTs norepinephrine (from tyrosine; see Figure 4.8) and serotonin (from tryptophan) are synthesized, with probable functional implications (Figure 4.18) (Song et al., 2008; Arcego et al., 2020).

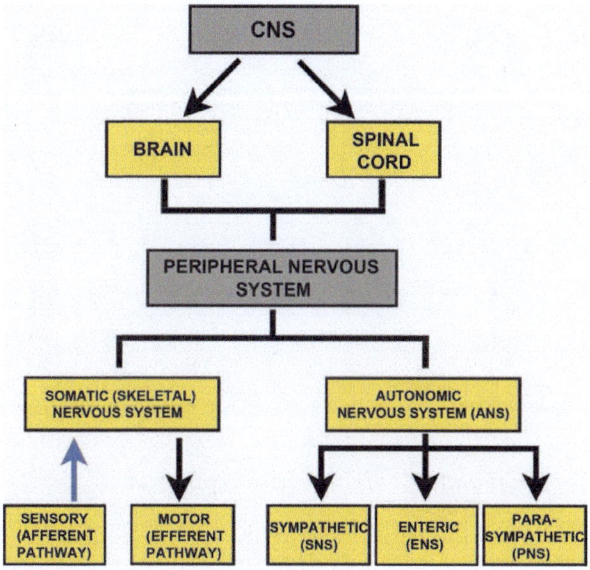

Figure 4.18. *The nervous system's divisions.*

Source: https://www.psychologyhub.co.uk/the-divisions-of-the-nervous-system-central-and-peripheral-somatic-and-autonomic/.

4.12. DIVISIONS OF THE NERVOUS SYSTEM

As illustrated in Figure 4.18, the nervous system is split into various divisions and sub-divisions. The CNS is made up of neurons and neural passageways in the brain, whereas the peripheral nervous system (PNS) is made up of motor nerves and peripheral sensory that branch out from the CNS and run throughout the body (Mattiasen et al., 2020). There are two divisions in the PNS: autonomic (autonomic) and somatic (skeletal).

4.13. THE SOMATIC NERVOUS SYSTEM

Through the spinal nerves and cranial nerves, the somatic division of the PNS receives sensory (afferent) input from the sense organs and regulates motor (efferent) output to the skeletal muscles. The skin and muscles send tactile information to the somatic sensory nerves, whereas the somatic motor nerves produce neuromuscular synapses that employ ACh as a NT. The somatic nervous system is responsible for conscious (voluntary) actions and reflexes, as well as sending motor nerves to muscles under voluntary control (Figure 4.19).

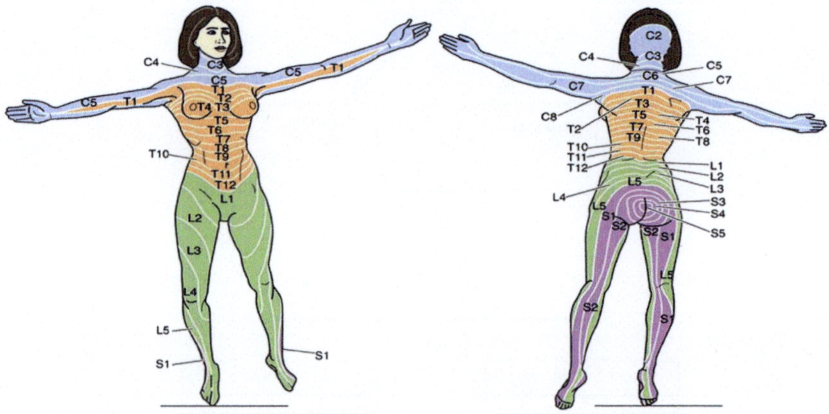

Figure 4.19. *Location and distribution of spinal nerves.*

Source: Reproduced with permission (Boron and Boulpaep, 2005); https://www.us.elsevierhealth.com/medical-physiology-9781455743773.html.

The optic nerve (vision), the olfactory nerve (smell), the acoustic nerve (hearing and balance), trochlear, the oculomotor, and abducens nerves (eye movement), the trigeminal nerve (controls jaw muscles and receives facial sensations), the facial nerve (receives taste stimuli and controls facial muscles), the glossopharyngeal nerve (receives taste stimuli and controls muscles of the larynx utilized in tongue), the vagus nerve (supplies heart, stomach, lungs, intestines, and kidneys), the accessory nerve (controls movement of head and neck) and the hypoglossal nerve (controls the movements of tongue).

4.14. THE AUTONOMIC NERVOUS SYSTEM (ANS)

In Figure 4.20, we see that the ANS receives sensory input from the viscera and sends motor nerves to the visceral (smooth) muscles of the heart, arteries, gastrointestinal system, and other organs, as well as to glandular tissues of the body, like the salivary glands, sweat glands, lachrymal (tear) glands, and peripheral endocrine glands (adrenal, thymus, gonads, and thyroid).

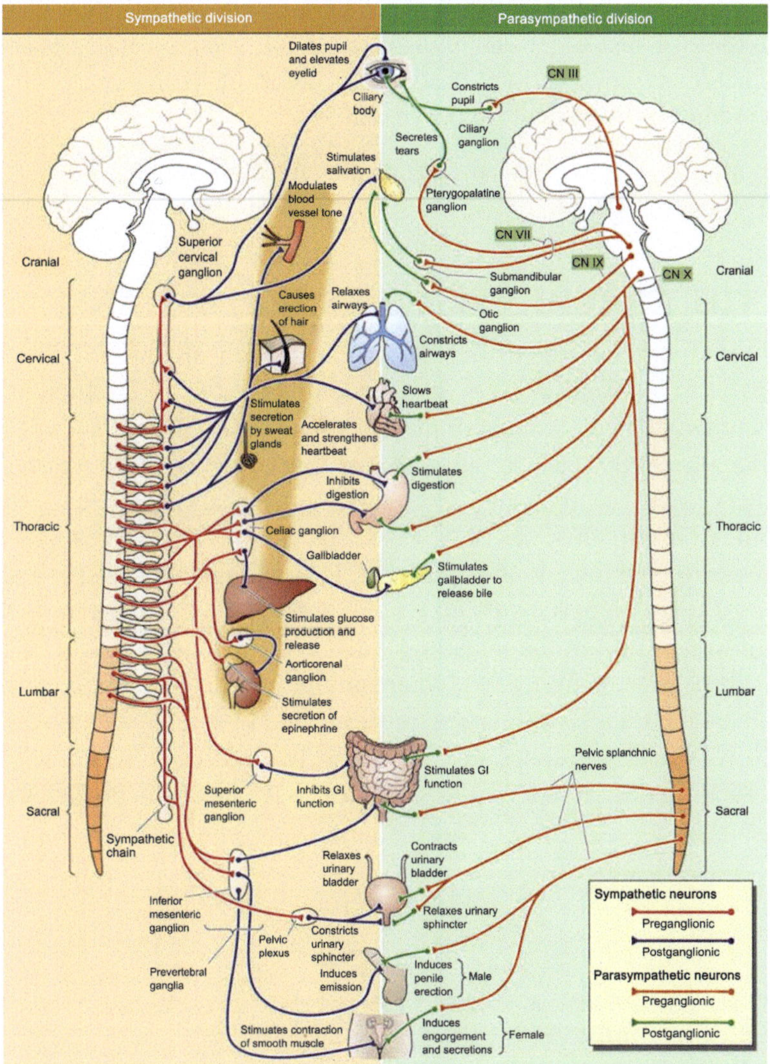

Figure 4.20. *Organization of the parasympathetic and sympathetic divisions of the autonomic nervous system (ANS).*

Note: The sympathetic division is seen in the left panel. The cell bodies of sympathetic preganglionic neurons (red) are found in the lumbar and thoracic spinal cord's intermediolateral column (T1-L3). Paravertebral ganglia (the sympathetic chain) and prevertebral ganglia get their axons. As a result, postganglionic neurons (blue) project enormous distances to their destinations. The parasympathetic division is seen on the right panel. Parasympathetic preganglionic neurons (orange) have cell bodies in either the brain (medulla, pons, midbrain) or the sacral spinal cord (S2-S4). Their axons branch out to ganglia that are located close to (or even inside) the end organs. As a result, the projections of postganglionic neurons (green) to their destinations are short. With permission, it has been reproduced (Boron and Boulpaep, 2005).

Source: https://www.us.elsevierhealth.com/medical-physiology-9781455743773.html.

4.15. THE ENTERIC NERVOUS SYSTEM (ENS)

The enteric nervous system (ENS), the 3rd branch of the ANS, regulates intestinal secretion and motility. It is made up of about 100 million neurons that are organized into two nerve plexuses within the GI tract's wall (Figure 4.21; Benarroch, 2007). Even though it is innervated by CNS neurons, this small nerve network may function autonomously and has been dubbed a mini-brain. Despite this, the PNS, ENS, and SNS as a whole exercise tight control over the GI tract, with neural signals traveling both ways, to and from the CNS (Figure 4.22) (Goyal and Hirano, 1996). The ENS's sensory nerves relay information from the intestinal walls and monitor the gastrointestinal tract's chemical environment, whereas the motor neurons govern the gastrointestinal tract's vasculature, muscles, and glandular secretions. More than 20 NTs and neuropeptides, including GABA, catecholamines, nitric oxide, ATP, peptide, and opioid peptides, have been identified as ENS components.

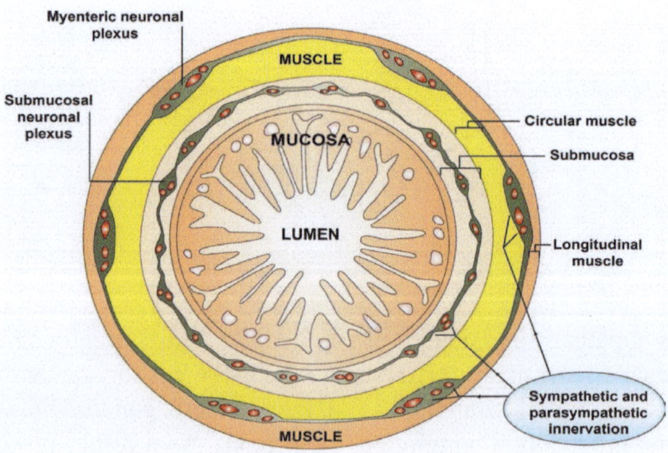

Figure 4.21. *The enteric nervous system.*

Note: A transverse segment through the small intestine is depicted schematically. Enteric neurons are arranged into ganglia (green) that are contained within two major plexi. Between the circular and longitudinal muscle layers are where the outer myenteric plexus is located. Within the submucosa, there is an inner submucosal plexus. The enteric nervous system is associated with the central nervous system (CNS) through ANS neurons. Permission is required for reproduction (Heanue and Pachnis, 2007).

Source: https://www.nature.com/articles/nrn2137.

Galanin and YY, along with these signaling molecules, endocrine cells in the GI tract and stomach secrete a large number of other hormones that function as paracrine mediators.

4.16. CENTRAL REGULATION OF THE AUTONOMIC NERVOUS SYSTEM (ANS)

The CNS controls all three branches of the ANS, as indicated in Figure 4.20. Through the nucleus of the solitary tract, which transmits fibers to the brainstem nuclei and spinal autonomic ganglia, hypothalamic, and limbic system nuclei govern the efferent passageways of the PNS, SNS, and ENS (shown in Figure 4.15). Visceral afferents from the periphery extend to the autonomic ganglia, which transmit fibers to the brain stem nuclei and spinal cord, as well as to the hypothalamus and limbic nuclei, as indicated in Figure 4.15. As a result, the hypothalamus combines ANS and neuroendocrine system activity (Boron and Boulpaep, 2005).

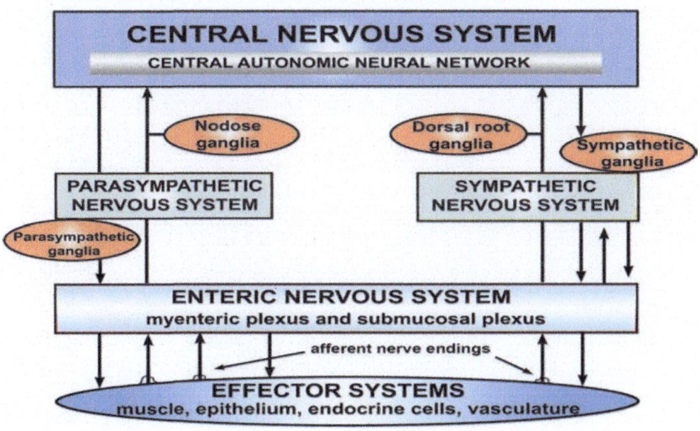

Figure 4.22. *Diagram of innervation of the GI tract.*

Note: The gut nerve plexuses are a self-contained network of nerves. The enteric nervous system (ENS) may either indirectly or directly govern the gut's effectors systems (such as muscle) via acting on intermediate cells, which include endocrine cells. Information travels in both directions, from the CNS to the ENS and vice versa. The vagal and splanchnic afferent neurons have their cell bodies in the nodose and dorsal root ganglia, respectively (Goyal and Hirano, 1996).

Source: https://pubmed.ncbi.nlm.nih.gov/8598871/.

REFERENCES

1. Altschuler, R. A., & Shore, S. E., (2010). Central auditory neurotransmitters. *The Oxford Handbook of Auditory Neuroscience, 2*, 65–89.
2. Ando, K., Ueyoshi, K., Orimo, K., Yonekawa, H., Sato, S., Nakahara, H., & Motomura, M., (2017). BRein memory: A 13-layer 4.2 K neuron/0.8 M synapse binary/ternary reconfigurable in-memory deep neural-network accelerator in 65 nm CMOS. In: *2017 Symposium on VLSI Circuits* (pp. C24, C25). IEEE.
3. Arcego, D. M., Krolow, R., Lampert, C., Toniazzo, A. P., Dos, S. G. E., Lazzaretti, C., & Dalmaz, C., (2020). Chronic high-fat diet affects food-motivated behavior and hedonic systems in the nucleus accumbens of male rats. *Appetite, 153*, 104739.
4. Baranano, D. E., Ferris, C. D., & Snyder, S. H., (2001). "Atypical neural messengers. *Trends Neurosci., 24,* 99–106.
5. Bartolome, J., Trepanier, P., Chait, E. A., Seidler, F. J., Deskin, R., & Slotkin, T. A., (1982). Neonatal methylmercury poisoning in the rat: Effects on development of central catecholamine neurotransmitter systems. *Toxicology and Applied Pharmacology, 65*(1), 92–99.
6. Belelli, D., & Lambert, J. J., (2005). "Neurosteroids: Endogenous regulators of the $GABA_A$ receptor." *Nature Rev. Neurosci., 6*, 565–575.
7. Benarroch, E. E., (2007). Enteric nervous system: Functional organization and neurologic implications. *Neurol., 69*, 1953–1957.
8. Besser, G. M., & Thorner, M. O., (2002). *Comprehensive Clinical Endocrinology* (3rd edn., Vol. 3, pp. 1–22). St. Louis, MO: Mosby.
9. Bolger, G. B., (2017). The PDE4 cAMP-specific phosphodiesterases: Targets for drugs with antidepressant and memory-enhancing action. *Phosphodiesterases: CNS Functions and Diseases*, 63–102.
10. Boron, W. F., & Boulpaep, E. L., (2005). *Medical Physiology* (Vol. 1, pp. 2–20). Philadelphia, PA: Elsevier Saunders.
11. Brady, S. T., Siegel, G. J., Albers, R. W., & Price, D. L., (2012). *Basic Neurochemistry* (8th edn., Vol. 3, pp. 1–22). New York: Elsevier.
12. Brann, D. W., (1995). "Glutamate: A major excitatory transmitter in neuroendocrine regulation." *Neuroendocrinology, 61*, 213–225.
13. Brann, D. W., Bhat, G. K., Lamar, C. A., & Mahesh, V. B., (1997). "Gaseous transmitters and neuroendocrine regulation." *Neuroendocrinology, 65*, 385–395.
14. Breivogel, C. S., & Sim-Selley, L. J., (2009). "Basic neuroanatomy and neuropharmacology of cannabinoids." *Int. Rev. Psych., 21*, 113–121.

15. Burgoyne, R. D., & Morgan, A., (2003). "Secretory granule exocytosis." *Physiol. Rev., 83*, 581–632.
16. Bustamante, D., You, Z. B., Castel, M. N., Johansson, S., Goiny, M., Terenius, L., & Herrera-Marschitz, M., (2002). Effect of single and repeated methamphetamine treatment on neurotransmitter release in substantia nigra and neostriatum of the rat. *Journal of Neurochemistry, 83*(3), 645–654.
17. Choi, S., Disilvio, B., Fernstrom, M. H., & Fernstrom, J. D., (2009). "Meal ingestion, amino acids and brain neurotransmitters: Effects of dietary protein source on serotonin and catecholamine synthesis rates." *Physiol, Behav., 98*, 156–162.
18. Christensen, R., Kristensen, P. K., Bartels, E. M., Bliddal, H., & Astrup, A., (2007). "Efficacy and safety of the weight-loss drug rimonabant: A meta-analysis of randomized trials." *Lancet, 370*, 1706–1713.
19. Christie, M. J., & Vaughn, C. W., (2001). "Cannabinoids act backwards." *Nature, 410*, 527–530. Crockett, M. J., Clark, L., Roiser, J. P., Robinson, O. J., Cools, R., Chase, H. W., et al., (2012). "Converging evidence for central 5-HT effects in acute tryptophan depletion." *Mol. Psych., 17*, 121–123.
20. Dahlin, M., Mansson, J. E., & Amark, P., (2012). "CSF levels of dopamine and serotonin, but not norepinephrine, metabolites are influenced by the ketogenic diet in children with epilepsy." *Epilepsy. Res., 99*, 132–138.
21. De Kloet, A. D., & Woods, S. C., (2009). "Minireview: Endocannabinoids and their receptors as targets for obesity." *Endocrinology, 150*, 2531–2536.
22. Deng, J., Lei, C., Chen, Y., Fang, Z., Yang, Q., Zhang, H., et al., (2014). "Neuroprotective gases – fantasy or reality for clinical use?" *Prog. Neurobiol., 115*, 210–245.
23. Di Guilmi, M. N., Urbano, F. J., Inchauspe, C. G., & Uchitel, O. D., (2011). Pregabalin modulation of neurotransmitter release is mediated by change in intrinsic activation/inactivation properties of Cav2. 1 calcium channels. *Journal of Pharmacology and Experimental Therapeutics, 336*(3), 973–982.
24. Eckstein, N., Bates, A. S., Du, M., Hartenstein, V., Jefferis, G. S., & Funke, J., (2020). Neurotransmitter classification from electron microscopy images at synaptic sites in *Drosophila*. *BioRxiv* (Vol. 1, pp. 1–22).
25. Eyigor, O., Centers, A., & Jennes, L., (2001). "Distribution of ionotropic glutamate receptor subunit mRNAs in the rat hypothalamus." *J. Comp. Neurol., 434*, 101–124.
26. Fernstrom, J. D., & Fernstrom, M. H., (2007). "Tyrosine, phenylalanine,

and catecholamine synthesis and function in the brain." *J. Nutr., 137*, 1539S–1547S. Discussion 1548S.

27. Fernstrom, J. D., (2012). "Large neutral amino acids: Dietary effects on brain neurochemistry and function." *Amino Acids, 45*, 419–430.
28. Fowler, C. J., Hall, H., Sääf, J., Ask, A. L., & Ross, S. B., (1985). Pharmacological manipulation of biochemically measured rhythms in the mammalian central nervous system. In: *Circadian Rhythms in the Central Nervous System* (Vol. 1, pp. 111–121). Palgrave Macmillan, London.
29. Gallagher, J. P., Zheng, F., Hasuo, H., & Shinnick-Gallagher, P., (1995). Activities of neurons within the rat dorsolateral septal nucleus (DLSN). *Progress in Neurobiology, 45*(5), 373–395.
30. Gerra, G., Zaimovic, A., Avanzini, P., Chittolini, B., Giucastro, G., Caccavari, R., & Brambilla, F., (1997). Neurotransmitter-neuroendocrine responses to experimentally induced aggression in humans: Influence of personality variable. *Psychiatry Research, 66*(1), 33–43.
31. Goyal, R. K., & Hirano, I., (1996). "The enteric nervous system." *N. Engl. J. Med., 334*, 1106–1115.
32. Groc, L., Choquet, D., Stephenson, F. A., Verrier, D., Manzoni, O. J., & Chavis, P., (2007). NMDA receptor surface trafficking and synaptic subunit composition are developmentally regulated by the extracellular matrix protein reelin. *Journal of Neuroscience, 27*(38), 10165–10175.
33. Guzmán, M., (2003). "Cannabinoids: Potential anticancer agents." *Nat. Rev. Cancer, 3*, 745–755.
34. Haas, H. L., Sergeeva, O. A., & Selbach, O., (2008). "Histamine in the nervous system." *Physiol. Rev., 88*, 1183–1241.
35. Heanue, T. A., & Pachnis, V., (2007). "Enteric nervous system development and Hirschsprung's disease: Advances in genetic and stem cell studies." *Nat. Rev. Neurosci., 8*, 466–479.
36. Hill, M. N., & Tasker, J. G., (2012). "Endocannabinoid signaling, glucocorticoid-mediated feedback and regulation of the hypothalamic-pituitary-adrenal axis." *Neuroscience, 204,* 5–16.
37. Iremonger, K. J., Constantin, S., Liu, X., & Herbison, A. E., (2010). "Glutamate control of GnRH neuron excitability." *Brain Res., 1364*, 35–43.
38. Iversen, L. L., Iversen, S. D., Bloom, D. E., & Roth, R. H., (2009). *Introduction to Neuropsychopharmacology* (Vol. 1, pp. 1–22). New York: Oxford University Press.
39. Jaen, J. C., Wise, L. D., Heffner, T. G., Pugsley, T. A., & Meltzer, L.

T., (1988). Dopamine autoreceptor agonists as potential antipsychotics. 1.(Aminoalkoxy) anilines. *Journal of Medicinal Chemistry, 31*(8), 1621–1625.

40. Jahan-Mihan, A., Luhovyy, B. L., El Khoury, D., & Anderson, G. H., (2011). "Dietary proteins as determinants of metabolic and physiologic functions of the gastrointestinal tract." *Nutrients, 3*, 574–603.
41. Jensen, A. A., Frølund, B., Liljefors, T., & Krogsgaard-Larsen, P., (2005). Neuronal nicotinic acetylcholine receptors: Structural revelations, target identifications, and therapeutic inspirations. *Journal of Medicinal Chemistry, 48*(15), 4705–4745.
42. Jorgensen, H. S., (2007). "Studies on the neuroendocrine role of serotonin." *Dan. Med. Bull., 54,* 266–288.
43. Julius, D., (1997). "Another opiate for the masses?" *Nature, 386*, 442.
44. Komsta, Ł., & Wicha-Komsta, K., (2018). Biological background of designer drugs. In: *Chromatographic Techniques in the Forensic Analysis of Designer Drugs* (pp. 39–66). CRC Press.
45. Kostoff, R. N., & Lau, C. G., (2013). Combined biological and health effects of electromagnetic fields and other agents in the published literature. *Technological Forecasting and Social Change, 80*(7), 1331–1349.
46. Krsmanovic, L. Z., Mores, N., Navarro, C. E., Saeed, S. A., Arora, K. K., & Catt, K. J., (1998). "Muscarinic regulation of intracellular signaling and neurosecretion in gonadotrophin-releasing hormone neurons." *Endocrinology, 139*, 4037–4043.
47. Kruk, Z. L., & Pycock, C. J., (1991). Dopamine. In: *Neurotransmitters and Drugs* (pp. 87–115). Springer, Dordrecht.
48. Kuhar, M., De Souza, E. B., & Unnerstall, J. R., (1986). "Neurotransmitter receptor mapping by autoradiography and other methods." *Annu. Rev. Neurosci., 9*, 27–59.
49. Lee, B. W., Lee, H. Y., Heo, K. H., Cho, H. W., Hwang, M. S., Heo, I., & Hwang, E. H., (2018). Analysis of herbal medicine prescriptions for patients in an academic Korean medical hospital: A cross-sectional study of electronic medical records (2010–2013). *Chinese Journal of Integrative Medicine, 24*(6), 467–473.
50. Lee, J. S., & Park, Y. C., (2014). Stability and interconversion of acetylcholine conformers. *Bulletin of the Korean Chemical Society, 35*(10), 2911–2916.
51. Levy, B. H., & Tasker, J. G., (2012). "Synaptic regulation of the hypothalamic-pituitary-adrenal axis and its modulation by glucocorticoids and stress." *Front Cell Neurosci., 6*, 24–32.

52. Maccarrone, M., & Wenger, T., (2005). "Effects of cannabinoids on hypothalamic and reproductive function." *Hand Exp. Pharmacol., 168*, 555–571.

53. Mackie, K., (2005). "Distribution of cannabinoid receptors in the central and peripheral nervous system," *Handb. Exp. Pharmacol., 168*, 299–325.

54. Madhavaiah, C., & Verma, S., (2005). Copper-metalated peptide palindrome derived from prion octarepeat: Synthesis, aggregation, and oxidative transformations. *Bioorganic & Medicinal Chemistry, 13*(9), 3241–3248.

55. Mansour, A., Fox, C. A., Akil, H., & Watson, S. J., (1995). "Opioid-receptor mRNA expression in the rat CNS: Anatomical and functional implications." *Trends Neurosci., 18*, 22–29.

56. Markus, C. R., Olivier, B., Panhuysen, E. M., Van, D. G. J., Alles, M. S., Tuiten, A., et al., (2000). "The bovine protein α-lactalbumin increases the plasma ratio of tryptophan to the other large neutral amino acids, and in vulnerable subjects raises brain serotonin activity, reduces cortisol concentration, and improves mood under stress." *Am. J. Clin. Nutr., 71*, 1536–1544.

57. Mattiasen, E. G., Kashef, N. S., Stafford, D. M., Logan, C. A., Sogard, S. M., Bjorkstedt, E. P., & Hamilton, S. L., (2020). Effects of hypoxia on the behavior and physiology of kelp forest fishes. *Global Change Biology, 26*(6), 3498–3511.

58. McCabe, B. J., (1986). Dietary tyramine and other pressor amines in MAOI regimens: A review. *Journal of the American Dietetic Association, 86*(8), 1059–1064.

59. Meadows, M. A., Balakrishnan, V., Wang, X., & Von, G. H., (2019). Regulation of neurotransmitter release during crossover inhibition. *Investigative Ophthalmology & Visual Science, 60*(9), 540–547.

60. Mehta, M. A., Gumaste, D., Montgomery, A. J., McTavish, S. F., & Grasby, P. M., (2005). "The effects of acute tyrosine and phenylalanine depletion on spatial working memory and planning in healthy volunteers are predicted by changes in striatal dopamine levels." *Psychopharmacology (Berl) 180*, 654–663.

61. Mendelsohn, D., Riedel, W. J., & Sambeth, A., (2009). "Effects of acute tryptophan depletion on memory, attention and executive functions: A systematic review." *Neurosci. Biobehav. Rev., 33*, 926–952.

62. Miller, K. K., (2011). "Endocrine dysregulation in anorexia nervosa update." *J. Clin. Endocr. Metab., 96,* 2939–2949.

63. Miller, L. L., Lau, A. N., & Miller, E. K., (1982). Electrically stimulated

release of neurotransmitters from a surface. An analog of the presynaptic terminal. *Journal of the American Chemical Society, 104*(19), 5242–5244.

64. Milner, J. D., & Wurtman, R. J., (1986). "Catecholamine synthesis: Physiological coupling to precursor supply." *Biochem. Pharmacol., 35*, 875–881.

65. Nestler, E. J., Hyman, S. E., & Malenka, R. C., (2009). *Molecular Neuropharmacology: A Foundation for Clinical Neuroscience* (2nd edn.). New York: McGraw-Hill.

66. Nicholls, J. G., Martin, A. R., Wallace, B. G., & Fuchs, P. A., (2001). *From Neuron to Brain* (4th edn., Vol. 1, pp. 1–26). Sunderland: Sinauer Associates.

67. Okuyama, S., Chaki, S., Yoshikawa, R., Suzuki, Y., Ogawa, S. I., Imagawa, Y., & Tomisawa, K., (1997). *In vitro* and *in vivo* characterization of the dopamine D4 receptor, serotonin 5-HT2A receptor and alpha-1 adrenoceptor antagonist (R)-(+)-2-amino-4-(4-fluorophenyl)-5-[1-[4-(4-fluorophenyl)-4-oxobutyl] pyrrolidin-3-yl] thiazole (NRA0045). *Journal of Pharmacology and Experimental Therapeutics, 282*(1), 56–63.

68. Olson, K. R., (2011). "The therapeutic potential of hydrogen sulfide: Separating hype from hope." Am. *J. Physiol. Regul. Integr. Comp. Physiol., 301*, R297–312.

69. Ong, M. L., & Holbrook, J. D., (2014). Novel region discovery method for Infinium 450K DNA methylation data reveals changes associated with aging in muscle and neuronal pathways. *Aging Cell, 13*(1), 142–155.

70. Opladen, T., Cortès-Saladelafont, E., Mastrangelo, M., Horvath, G., Pons, R., Lopez-Laso, E., & Garcia-Cazorla, À., (2016). The international working group on neurotransmitter related disorders (iNTD): A worldwide research project focused on primary and secondary neurotransmitter disorders. *Molecular Genetics and Metabolism Reports, 9*, 61–66.

71. Otto, M. W., (2011). "Expanding findings on D-cycloserine augmentation of therapeutic learning: A role for social learning relative to autism spectrum disorders?" *Biol. Psych., 70*, 210, 211.

72. Pagotto, U., Marsicano, G., Cota, D., Lutz, B., & Pasquali, R., (2006). "The emerging role of the endocannabinoid system in endocrine regulation and energy balance." *Endocr. Rev., 27*, 73–100.

73. Palmada, M., & Centelles, J. J., (1998). Excitatory amino acid neurotransmission. Pathways for metabolism, storage, and reuptake of glutamate in brain. *Front Biosci., 3*, d701–d718.

74. Papakostas, G. I., Miller, K. K., Petersen, T., Sklarsky, K. G., Hilliker, S. E., Klibanski, A., et al., (2006). "Serum prolactin levels among outpatients

with major depressive disorder during the acute phase of treatment with fluoxetine." *J. Clin. Psychiatry, 67*, 952–957.

75. Perry, E., Walker, M., Grace, J., & Perry, R., (1999). "Acetylcholine in mind: A neurotransmitter correlate of consciousness?" *Trends Neurosci., 22*, 273–280.
76. Prevot, V., Bouret, S., Stefano, G. B., & Beauvillain, J., (2000). "Median eminence nitric oxide signaling." *Brain Res. Rev., 34*, 27–41.
77. Pritchett, C. E., Pardee, A. L., McGuirk, S. R., & Will, M. J., (2010). The role of nucleus accumbens adenosine–opioid interaction in mediating palatable food intake. *Brain Research, 1306*, 85–92.
78. Purves, D., Augustine, G. J., Fitzpatrick, D., Hall, W. C., LaMantia, A. S., McNamara, J. O., et al., (2008). *Neuroscience* (4th edn., Vol. 4, pp. 1–22).
79. Racagni, G., Trabucchi, M., & Cheney, D. L., (1975). Steady-state concentrations of choline and acetylcholine in rat brain parts during a constant rate infusion of deuterated choline. *Naunyn-Schmiedeberg's Archives of Pharmacology, 290*(1), 99–105.
80. Rettori, V., De Laurentiis, A., & Fernandez-Solari, J., (2010). "Alcohol and endocannabinoids: Neuroendocrine interactions in the reproductive axis." *Exp. Neurol., 224*, 15–22.
81. Rhee, J. S., Li, L. Y., Shin, O. H., Rah, J. C., Rizo, J., Südhof, T. C., & Rosenmund, C., (2005). Augmenting neurotransmitter release by enhancing the apparent Ca^{2+} affinity of synaptotagmin 1. *Proceedings of the National Academy of Sciences, 102*(51), 18664–18669.
82. Richard, D. M., Dawes, M. A., Mathias, C. W., Acheson, A., Hill-Kapturczak, N., & Dougherty, D. M., (2009). "L-tryptophan: Basic metabolic functions, behavioral research and therapeutic indications." *Int. J. Tryptophan Res., 2*, 45–60.
83. Riedel, W. J., Klaassen, T., & Schmitt, J. A., (2002). "Tryptophan, mood, and cognitive function." *Brain Behav. Immun., 16*, 581–589.
84. Rodrigo, J., Springall, D. R., Uttenthal, O., Bentura, M. L., Abadia-Molina, F., Riveros-Moreno, V., & Moncada, S., (1994). Localization of nitric oxide synthase in the adult rat brain. *Philosophical Transactions of the Royal Society of London. Series B: Biological Sciences, 345*(1312), 175–221.
85. Rose, M., Dütting, E., & Enz, R., (2008). Band 4.1 proteins are expressed in the retina and interact with both isoforms of the metabotropic glutamate receptor type 8. *Journal of Neurochemistry, 105*(6), 2375–2387.
86. Santos, M. S., Moreno, A. J., & Carvalho, A. P., (1996). Relationships

between ATP depletion, membrane potential, and the release of neurotransmitters in rat nerve terminals: An *in vitro* study under conditions that mimic anoxia, hypoglycemia, and ischemia. *Stroke, 27*(5), 941–950.

87. Sasse, B. C., (2007). *Development of Novel Ligands Influencing Neurotransmission in the Central Nervous System*. Doctoral dissertation.
88. Scheller, R. H., (2013). "In search of the molecular mechanism of intracellular membrane fusion and neurotransmitter release." *Nat. Med., 19*, 1232–1235.
89. Shiao, Y. J., Su, M. H., Lin, H. C., & Wu, C. R., (2017). Acteoside and isoacteoside protect amyloid β peptide-induced cytotoxicity, cognitive deficit and neurochemical disturbances *in vitro* and *in vivo*. *International Journal of Molecular Sciences, 18*(4), 895.
90. Song, C., Manku, M. S., & Horrobin, D. F., (2008). Long-chain polyunsaturated fatty acids modulate interleukin-1 β–induced changes in behavior, monoaminergic neurotransmitters, and brain inflammation in rats. *The Journal of Nutrition, 138*(5), 954–963.
91. Squire, L. R., Berg, D. E., Bloom, F. E., Du Lac, S., Ghosh, A., & Spitzer, N. C., (2008). *Fundamental Neuroscience* (3rd edn.). London: Academic Press.
92. Stanford, S. C., (2001). Four neurotransmitter release. *Neurotransmitters, Drugs and Brain Function,* 81.
93. Sullivan, J., (2003). "No going back." *Nat. Neurosci., 6*, 905, 906.
94. Sundvik, M., (2012). *The Histaminergic Neurotransmitter System: Its Development and Role in Neuronal Plasticity in Zebrafish, Danio Rerio.*
95. Suzuki, H., Yoshioka, K., Yanagisawa, M., Urayama, O., Kurihara, T., Hosoki, R., & Otsuka, M., (1994). Involvement of enzymatic degradation in the inactivation of tachykinin neurotransmitters in neonatal rat spinal cord. *British Journal of Pharmacology, 113*(1), 310–316.
96. Thio, L. L., (2012). "Hypothalamic hormones and metabolism." *Epilepsy Res., 100*, 245–251.
97. Tiwari, V., Veeraiah, P., Subramaniam, V., & Patel, A. B., (2014). Differential effects of ethanol on regional glutamatergic and GABAergic neurotransmitter pathways in mouse brain. *Journal of Neurochemistry, 128*(5), 628–640.
98. Und Halbach, O. V. B., & Dermietzel, R., (2006). *Neurotransmitters and Neuromodulators: Handbook of Receptors and Biological Effects*. John Wiley & Sons.
99. Van, D. P. A. N., Hermans-Borgmeyer, I., Hofer, M., Ghosh, P., &

Heinemann, S., (1994). "Ionotropic glutamate-receptor gene expression in hypothalamus: Localization of AMPA, kainate, and NMDA receptor RNA with in situ hybridization." *J. Comp. Neurol., 343*, 428–444.

100. Van, D. P. A. N., Wuarin, J. P., & Dudek, F. E., (1990). "Glutamate, the dominant excitatory transmitter in neuroendocrine regulation." *Science, 250,* 1276–1278.

101. Van, G. L. F., Rissanen, A. M., Scheen, A. J., Ziegler, O., & Rössner, S., (2005). "Effects of cannabinoid-1 receptor blocker rimonabant on weight reduction and cardiovascular risk factors in overweight patients: 1-year experience from the RIO-Europe study." *Lancet, 365*, 1389–1397.

102. Vincent, S. R., (2010). "Nitric oxide neurons and neurotransmission." *Prog. Neurobiol., 90*, 246–255.

103. Visser, G. P., (2012). *Cortical Brain Release of Glutamate by Ketamine and Fluoxetine: An in Vivo Microdialysis Study in the Flinders Sensitive Line Rat* (Vol. 1, pp. 1–21). Doctoral dissertation, North-West University.

104. Vizi, E. S., & Kiss, J. P., (1998). Neurochemistry and pharmacology of the major hippocampal transmitter systems: Synaptic and nonsynaptic interactions. *Hippocampus, 8*(6), 566–607.

105. Walz, W., (1985). Uptake and release of amino acid neurotransmitters. In: *Amino Acids* (Vol. 1, pp. 239–272). Humana Press.

106. Ward, S. J., & Raffa, R. B., (2011). "Rimonabant redux and strategies to improve the future outlook of CB1 receptor neutral-antagonist/inverse-agonist therapies." *Obesity, 19*, 1325–1334.

107. Watanabe, M., Inoue, Y., Sakimura, K., & Mishina, M., (1993). "Distinct distribution of five N-methyl-D aspartate receptor channel subunit mRNAs in the forebrain." *J. Comp. Neurol., 338*, 377–390.

108. Watkins, J. C., & Jane, D. E., (2006). "The glutamate story." *Brit. J. Pharmacol., 147*, S100–S108.

109. Wittman, G., Lechan, R. M., Liposits, Z., & Fekete, C., (2005). "Glutamatergic innervation of corticotropin-releasing hormone- and thyrotropin-releasing hormone-synthesizing neurons in the hypothalamic paraventricular nucleus of the rat." *Brain Res., 1039*, 53–62.

110. Wurtman, R. J., (1982). "Nutrients that modify brain function." *Sci. Am., 246*, 50–59.

111. Xue, W., Gao, Y., Xie, P. P., Liu, Y., Qi, W. Y., Shi, A. X., & Li, K. X., (2021). Plasma and intracerebral pharmacokinetics and pharmacodynamics modeling for the acetylcholine releasing effect of ginsenoside Rg1 in mPFC of A β model rats. *Journal of Asian Natural Products Research, 23*(3), 294–306.

112. Zador, A. M., Claiborne, B. J., Brown, T. H., & Clairborne, B. J., (1993). Nonlinear pattern separation in single hippocampal neurons with active dendritic membrane. *Advances in Neural Information Processing Systems, 1*, 51–55.

CHAPTER

5

Neurochemistry of Neuropeptides and Psychopharmaceuticals

Contents

5.1. INTRODUCTION

When compared to neurotransmitter (NT) receptors, neuropeptide signals have the potential to be more distinct than signals produced by more widely expressed conventional NTs, as well as lower molecular weight. As a result, active pharmaceutical ingredients taking up neuropeptide receptors have highly selective pharmacological action and consequently fewer side effects when compared to active molecules occupying NT receptors. A large number of neuropeptides and their corresponding receptors opens the door to the development of novel therapeutic targets for illnesses of the central nervous system (CNS). All agonist and antagonist subtypes of selective receptors, as well as modulators of neuropeptide receptors, are being researched and produced. The usefulness of these ligands has been demonstrated in both preclinical and clinical models of pain, as well as in neuropsychiatric disorders. In this chapter, we will explore how neuropeptide neurochemicals and psychopharmaceuticals alter the function of neural, synapses, and neurons networks by binding to various receptors and causing changes in brain-behavioral states.

The study of how neurochemicals such as neuropeptides, psychopharmaceuticals, and NTs impact the activity of neurons, synapses, and neural networks is referred to as neurochemical chemistry or neurochemical biology. Some changes in the electrophysiological activity of these neurochemicals have the potential to cause changes in the brain and in behavioral situations (de Girolamo and Dieguez, 2014; Shuler and Hickman, 2014). For example, the neuropeptide ghrelin, which is expressed in both the brain and the stomach, stimulates food intake and energy balance. The neuropeptide VF (NPVF) (vertebrate hypothalamic RFamide neuropeptide VF) is involved in the regulation of sleep (Lee et al., 2017). Psychosis develops thus of an imbalance in dopaminergic activity in the brain (Tost et al., 2010).

5.2. NEUROPEPTIDES

"Neuropeptides are tiny polypeptides that are released and generated by neurons via the controlled secretory pathway, behaving on neural substrates as a result (Burbach, 2011). They are a sort of neurochemical that has an effect on both the activity of the brain's neuronal system and the activity of the peripheral nervous system (PNS). Neuropeptides are key components in the brain, as they regulate behaviors such as memory and other cognitive functions.

When it comes to the production of peptides, the neuron is the most important factor to consider (Russo, 2017). Glial cells, on the other hand, produce neuropeptides (Hökfelt et al., 2000).

5.3. CLASSIFICATION OF NEUROPEPTIDE

The number of neuropeptides encoded by the human genome has topped 250 molecules, together with over 100 molecules found in the CNS. On the basis of structural homology, neuropeptides are divided into several families (Wang et al., 2015).

As previously stated, neuropeptides are found in a variety of CNS neurons and affect neuronal activities (Burbach, 2011).

5.4. SECRETION OF NEUROPEPTIDE

Neuropeptides are released as dense-core vesicles (DCVs) alongside the NT's tiny vesicles or as dense basic vesicles on their own. This is largely seen in the hypothalamus; nevertheless, GABAergic neurons in the hippocampus area produce a plethora of neuropeptides, including substance P, neuropeptide Y, enkephalin, vasoactive intestinal polypeptide (VIP), cholecystokinin (CCK), Somatostatin, dynorphin, and neurokinin B (van den Pol, 2012). The information below concentrates on the functions of a few of the most well-studied neuropeptides.

5.4.1. Apelin

Apelin is a protein that interacts with the APJ receptor and is encoded by the APLN gene. Adipose tissue, the gastrointestinal system, the brain, the heart, the lungs, the kidneys, and the liver are all affected (Tatemoto et al., 1998; Wysocka et al., 2018). They are characterized as apelin 13, apelin 17, and apelin 36 after cleavage from proprotein. Appetite and water intake are controlled by apelin expression in the brain (De Mota et al., 2000; Sunter et al., 2003). It reduces the antidiuretic hormone vasopressin production in the hypothalamus. The diuretic impact of this, together with its hypotensive effect, aids in the maintenance of body fluid balance. It also aids in appetite management (O'Carroll et al., 2000; Taheri et al., 2002). Figure 5.1 shows the structural formula of apelin.

5.4.2. Neuropeptide Y

Together with the GABA and neurochemicals glutamate, it is the most extensively released neuropeptide in the human CNS. Stress response, circadian rhythms, food intake, cortical excitability, and cardiovascular function are all affected by neuropeptides. Neuropeptide Y also affects sleep by regulating noradrenergic signaling (Singh et al., 2017; Diaz-Del et al., 2018). Higher alcohol intake, higher cholesterol levels, and a higher risk of metabolic and cardiovascular illnesses are all linked to changes in gene activity (Watkins et al., 2017; Melas et al., 2018). Figure 5.2 shows the structural formula of neuropeptide Y.

5.4.3. Neuromedin

Neuromedin U is a neuropeptide found in the gastrointestinal system and pituitary gland. Smooth muscle contraction, blood pressure, blood flow, intestinal ion movement, tumor, stomach acid secretion stress regulation, eating activities, and pronociceptive have all been linked to it (Brighton et al., 2004; Martinez et al., 2014).

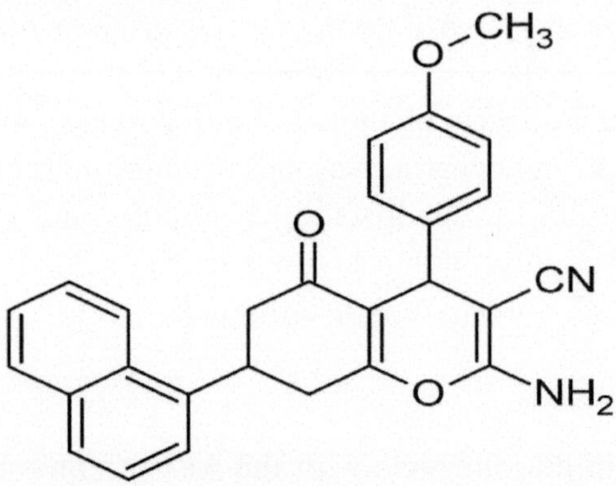

Figure 5.1. *Chemical structure of apelin.*

Source: https://www.abcam.com/pyr1-apelin-13-endogenous-apelin-receptor-apj-agonist-ab141012.html.

An additional bombesin-related peptide found in the gastrointestinal and CNS system is Neuromedin B. Its interaction with 7-TMR, a heterotrimeric G-protein, causes cyclic AMP response element-binding protein to be activated, which affects learning and long-term memory (Ohki-Hamazaki, 2000). Figure 5.3 shows the neuromedin structural formula.

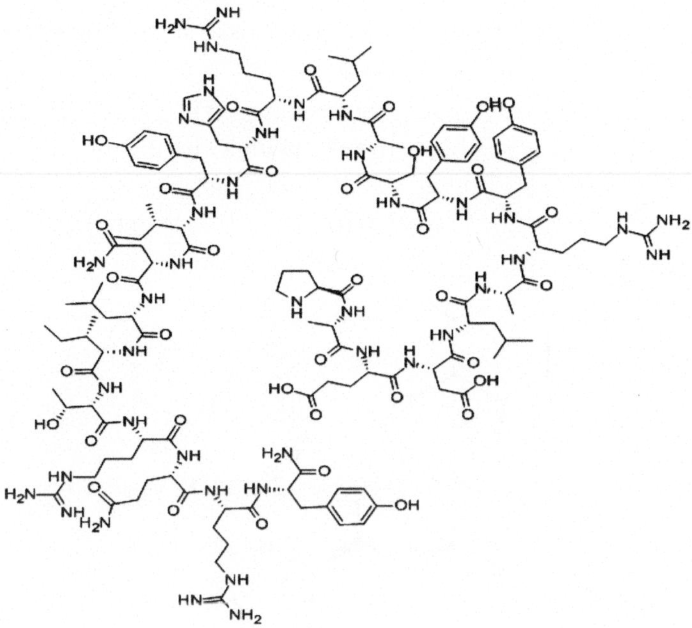

Figure 5.2. *The structural formula of neuropeptide Y.*

Source: https://www.chemicalbook.com/ChemicalProductProperty_EN_CB9271630.htm.

5.4.4. Enkephalins

It relates to the endorphin-like neuropeptide family. The opioid growth factor is another name for enkephalins. Figure 5.4 shows the chemical differences between the two forms of enkephalin. Met-enkephalin causes depressive and analgesic effects through binding to the -opioid receptor, and to a lesser amount, the -opioid receptor (Clement-Jones et al., 1980). It has also been discovered to affect human sperm motility. L enkephalin, which has a high affinity for the -opioid receptor and the -opioid receptor, has the same effect (Subirán et al., 2012).

5.4.5. Orexin

Orexin, also known as hypocretin, is a peptide that regulates hunger, arousal, and alertness, as well as motivated behavior (Inutsuka and Yamanaka, 2013; Sakurai, 2014). Chronic drowsiness is also caused by the loss of orexin-producing neurons in narcolepsy, an autoimmune condition (Mahoney et al., 2019). Orexin-A and orexin-B are two isoforms of the neuropeptide orexin. Figure 5.5 shows the structural formula of orexin.

Orexin-A creates neural signals by binding to OX1 and OX2 with almost equal affinity, but orexin-B binds to OX2 with a significant affinity. Orexin has also been linked to drug addiction, ischemic stroke, and Alzheimer's disease (AD) pathogenesis (Yamanaka et al., 2010; Wang et al., 2018).

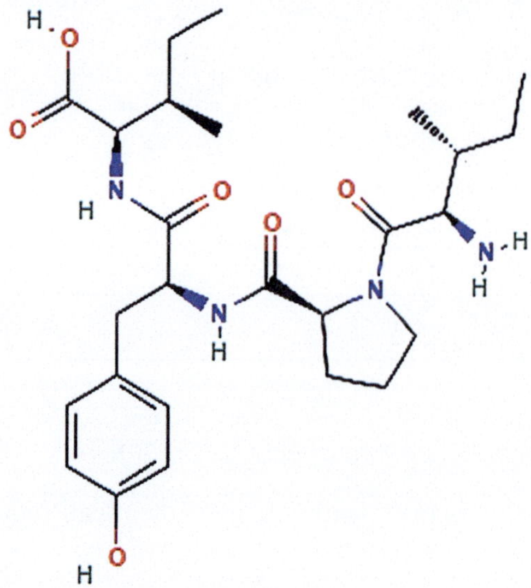

Figure 5.3. *The structural formula of neuromedin.*

Source: https://pubchem.ncbi.nlm.nih.gov/compound/Neuromedin-N-_1-4.

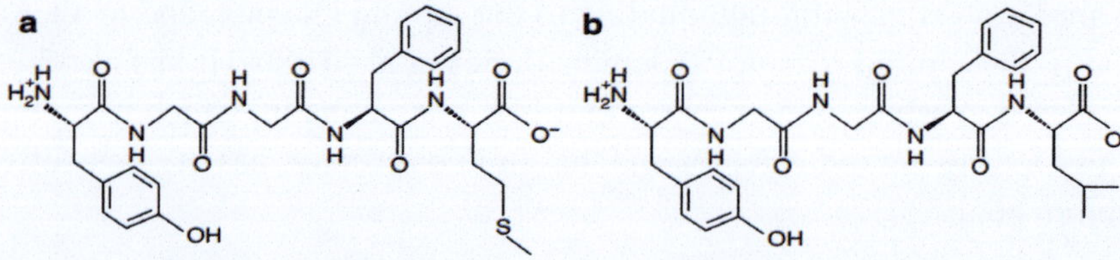

Figure 5.4. *The structural formula of (a) met-enkephalin; and (b) leu-enkephalin.*

Source: https://www.sciencedirect.com/science/article/abs/pii/S0003269718302562.

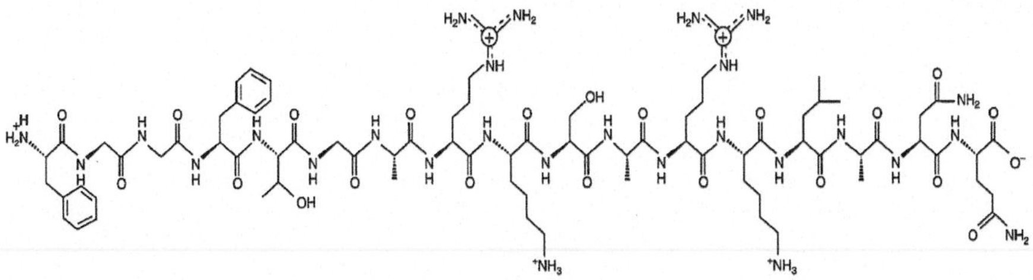

Figure 5.5. *The structural formula of orexin-B.*

Source: https://pubchem.ncbi.nlm.nih.gov/compound/Orexin-B-_human.

5.4.6. Nociceptin

Nociceptin, a neuropeptide of 17 amino acids found in the brain, is a ligand for the opioid receptor (Toll, 2016). Nociceptin is broadly dispersed throughout the body, allowing endogenic chemicals to influence a variety of physiological processes through it. Nociceptin is found in several regions of the CNS, including the forebrain, spinal cord, hypothalamus, and brainstem (Calo' et al., 2000). The bonding of FQ (N/OFQ), a neuropeptide of 17 amino acids, to the nociceptin receptor (NOP, ORL-1) opens the door to fear and pain learning in the brain, according to research. Nociceptin has the ability to counteract the effects of pain medications, resulting in anti-analgesic activity (Meunier et al., 1995; Okuda-Ashitaka et al., 1998). The binding of nociceptin enhances the threshold for pain and also diminishes acceptance to analgesic opioids (Zaveri, 2011), which are both beneficial. On the right is a representation of the structural formula of nociceptin (see Figure 5.6).

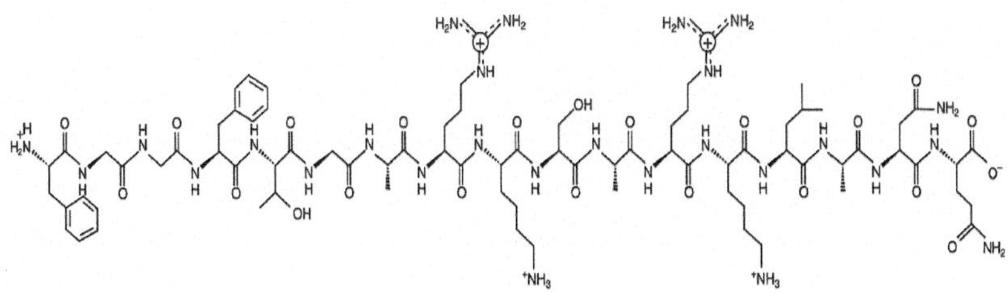

Figure 5.6. *The structural formula of nociception.*

Source: https://pubchem.ncbi.nlm.nih.gov/compound/Nociceptin.

Several animal investigations have demonstrated that nociceptin has a function in anxiolysis and depressive apprehension (Jenck et al., 1997; Gavioli et al., 2004).

5.4.7. Endothelin

It is the 21-amino-acid endothelin peptide that has been incorporated into the endothelium to constrict blood vessels. They are involved in the development of vascular disorders in a variety of organ systems, including the kidney, lung, heart, and brain (Agapitov and Haynes, 2002; Schinelli, 2006).

Endothelin is composed of three isoforms: ET 1, ET 2, and ET 3. These isoforms bind to the receptors ETA, ETB1, ETB2, and ETC. When endothelin-1 from the blood or the ventricular systems interacts with endothelin receptors in the pituitary lobe, the metabolic activity of the pituitary lobe is boosted. In Figure 5.7, we can see the structural formula of endothelin.

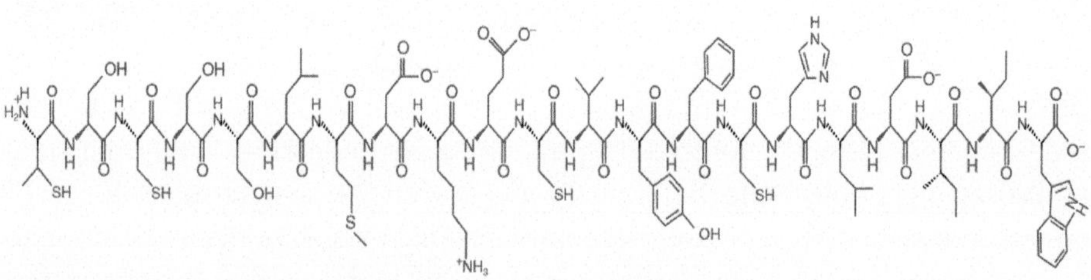

Figure 5.7. *The structural formula of endothelin.*

Source: https://www.caymanchem.com/product/24127/endothelin-1-(human%2C-porcine)-(trifluoroacetate-salt).

ET-1 levels in the cerebral spinal fluid were shown to be higher 18 hours after a stroke, and this had an effect on the patient's neurological symptoms (Kuhn et al., 2006).

5.4.8. Prolactin-Releasing Peptide

This neuropeptide interacts with the PrRP receptor and regulates endocrine function and energy metabolism. The amino acids PrRP20 and PrRP31 are two physiologically active versions of receptors. The phosphorylation of MAPK/ERK1/2. It has also been shown to lower food consumption, making it useful in the treatment of obesity (Maixnerová et al., 2011; Tachibana and Sakamoto, 2014). Figure 5.8 shows the structural formula of the prolactin-releasing peptide.

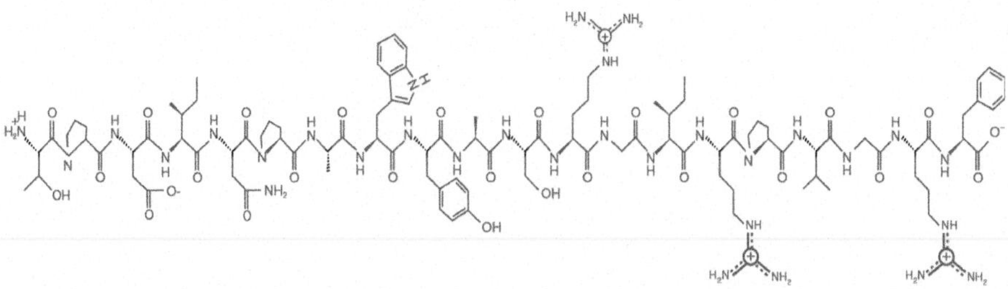

Figure 5.8. *The structural formula of prolactin-releasing peptide.*

Source: https://www.mybiosource.com/prlh-peptide/prolactin-releasing-peptide/8246977.

5.4.9. Galanin

Galanin can be found in the spinal cord, brain, and intestines of humans and other animals. The GAL gene controls it, and three GPCRs signal it (Evans et al., 1993; Mitsukawa et al., 2008).

Galanin's main function is to modulate and inhibit neuronal activity potentials. Nociception, cognition, eating, mood management, blood pressure regulation, sleep, and waking regulation, along with serving as a trophic factor, are all activities of galanin. AD, depression, epilepsy, cancer, and eating disorders are among the ailments linked to it (Lundström et al., 2005). Galanin has neuroprotective properties, as evidenced by an increase in its production during brain convulsions. It also has a role in neurogenesis.

Galanin is a hyperpolarizing neuropeptide that suppresses NT release. Galanin interacts with NTs such as norepinephrine, serotonin, and acetylcholine (ACh), along with neuromodulators such vasoactive intestinal peptide, substance P, and neuropeptide Y. Figure 5.9 shows the galanin structural formula.

5.4.10. Neurotensin

Neurotensin is a 13-amino-acid peptide discovered in the CNS that has a role in the digestive system. The action of the agonists and endogenous peptides is mediated via the NT-1 and NT-2 receptors.

It includes analgesia, thermoregulation, and connection with dopamine (DA) function in the nigrostriatal and mesolimbic pathways (St-Gelais et al., 2006). It is thought to play a role in schizophrenia, drug addiction, Parkinson's disease, pain, eating disorders, cancer, and inflammation. Its structural formula is shown in Figure 5.10.

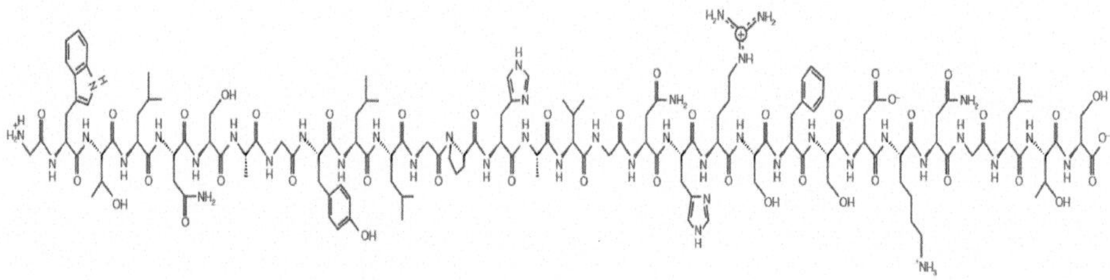

Figure 5.9. *The structural formula of galanin.*

Source: https://www.chemicalbook.com/ChemicalProductProperty_EN_CB9732322.htm.

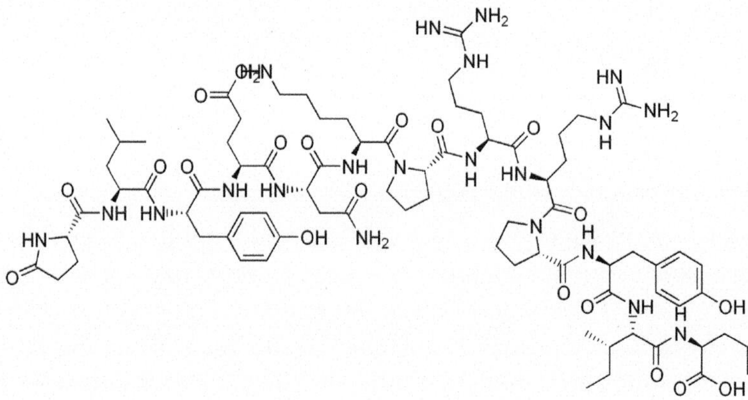

Figure 5.10. *The structural formula of neurotensin.*

Source: https://en.wikipedia.org/wiki/Neurotensin.

5.4.11. Vasopressin

Vasopressin has been linked to cognitive improvement and memory learning in the CNS. Figure 5.11 shows the structural formula of vasopressin.

V1a, V1b, and V2 receptors are involved in its activity (Alescio and Soumireu, 1998). V1a and Vasopressin receptors are implicated in maintaining ion homeostasis and brain water, which is expected to be accomplished by aquaporin-mediated water transport across astrocyte plasma membranes (Niermann, 2001).

5.4.12. Somatostatin

Somatostatin is found in two different isoforms: SRIF 14 and SRIF 28. They are widely distributed throughout the CNS and have a function in activating potassium channels, resulting in an inhibitory action on neurons (Córdoba et al., 2012).

It is SRIF 14 that is responsible for action by binding to sst 2, 3, and 4, whereas SRIF 28 is responsible for action by binding to sst 1, 2, 3, and 4. Antinociception is produced in the spinal cord by this NT, which is also involved in the regulation of the release of growth hormone. On the right-hand side of the figure, you can see the structural formula of Somatostatin.

Among GABA, the existence of somatostatin in dendritic targeting GABA neuron loss contributes to cortical local inhibitory circuit deficit, which results in aberrant cortex limbic network activity and subjective mood symptoms crosswise a range of neurological disorders (Figure 5.12) (Lin and Sibille, 2013).

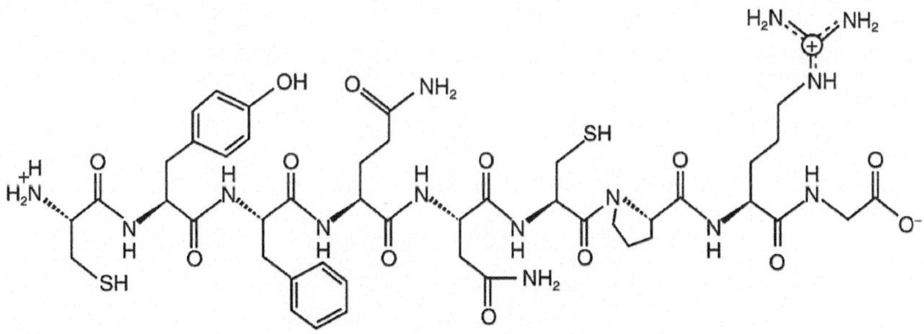

Figure 5.11. *The structural formula of vasopressin.*

Source: https://en.wikipedia.org/wiki/Vasopressin.

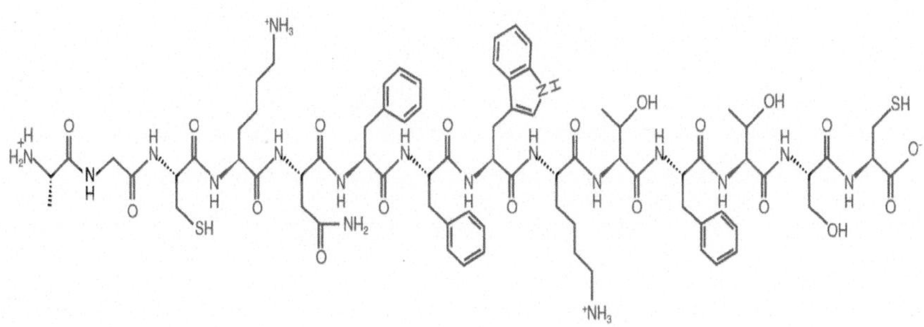

Figure 5.12. *The structural formula of somatostatin 14.*

Source: https://pubchem.ncbi.nlm.nih.gov/compound/Somatostatin-14.

5.5. RECEPTORS FOR NEUROPEPTIDE

Neuropeptides are neurochemicals that act on cell surface receptors to create neuronal impulses. Neuropeptide release is not limited to the synapse, as it is with NTs, but it can travel across a given distance after discharge to create its action through a receptor-mediated by G-protein. Activation causes the G-protein to transition from GDP to GTP, dissociating it from its associated receptor. The stimulation of various intracellular secondary signal-producing pathways, such as activation of adenylyl cyclase-cAMP and the phosphatidylinositol lane, augments further peptide signals. Peptides have been discovered on axon terminals, dendrites, and cell bodies in addition to postsynaptic receptors.

5.6. PSYCHOPHARMACEUTICALS

The term "psychopharmaceuticals" refers to any pharmacological agent that alters mood, perception, or consciousness as a result of changes in the brain neuronal network's function. As a result, psychopharmaceuticals are confined not only to the treatment of brain-related disorders nevertheless also to the treatment of cognitive changes (Kline, 1959).

5.6.1. Neurodegenerative Disorder

Neurodegenerative illnesses cause the neuron to break down or die. These illnesses characteristics are discussed in subsections.

5.6.1.1. Parkinson's Disease

- Loss or slowness of movement called akinesia or bradykinesia;
- Vibrations of limbs;
- Stiffness and rigidity of muscle.

Due to the degeneration of neurons in the substantia nigra, this is visible. Because it creates a dopaminergic nigrostriatal area, Parkinson's disease causes striatal DA depletion, which is a hallmark of the disease.

As a result, psychopharmaceuticals for therapy should concentrate on the subsequent:

- Enhance the action of DA;
- Alter the actions of other neurotransmitters (NTs) that may enhance or oppose DA (dopamine) effects.

This can be accomplished by boosting the activity of residual DA, replacing DA by providing levodopa, simulating the effect of DA receptor agonists, preventing the function of DA-blocking NTs, and preventing the routes that DA stimulates or inhibits.

1. **Increasing the Action of Remaining Dopamine (DA):** It can be accomplished by reducing DA reuptake in presynaptic neurons, such as with nomifensine, a powerful DA reuptake inhibitor, and by specifically inhibiting the monoamine oxygenase B enzymes, such as with safinamide, selegiline, and rasagiline.
2. **Replenishment of Dopamine (DA):** This impact cannot be obtained by administering DA intravenously, but it can be obtained by administering levodopa.

In the peripheral, the enzyme DOPA decarboxylase transforms levodopa to DA, which is then converted to 3-methoxy tyrosine by COMT. This can be avoided by using the extracerebral decarboxylase inhibitors benserazide and carbidopa. Entacapone and tolcapone are inhibitors of catecholamine methyltransferase (Jankovic, 2008). Figures 5.13–5.16 demonstrate the chemical structures of medications used to cure disorders.

Postsynaptically, DA agonists can be employed since the number of DA receptors does not diminish. Anti-parkinsonism action is produced by agonists for DA binding completely to DA receptors and therefore limiting endogenous NTs. These categories are focused on DA D2-type receptors.

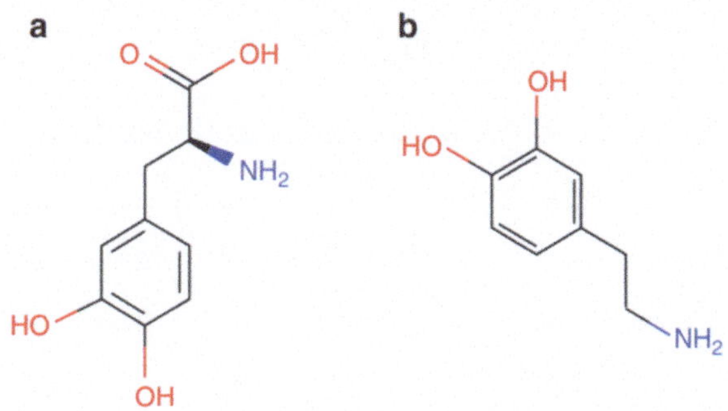

Figure 5.13. *The structural formula of (a) levodopa; and (b) dopamine.*

Source: https://www.researchgate.net/figure/Chemical-structures-of-the-neurotransmitter-dopamine-and-levodopa_fig5_272565621.

Apomorphine is a subcutaneously administered DA receptor agonist (Katzenschlager et al., 2018). DA inhibits cholinergic neuron (CN) activity as well as ACh release via the D2 receptor. Withdrawn ACh stimulates GABA/Enk neurons via M1 receptors, which counteract DA's inhibitory effect. The production of ACh is increased in the absence of DA, and its stimulating effect on Enk /GABA cells cannot be prevented. This permits the antimuscarinic medication atropine to be used (Stockbrügger, 1989).

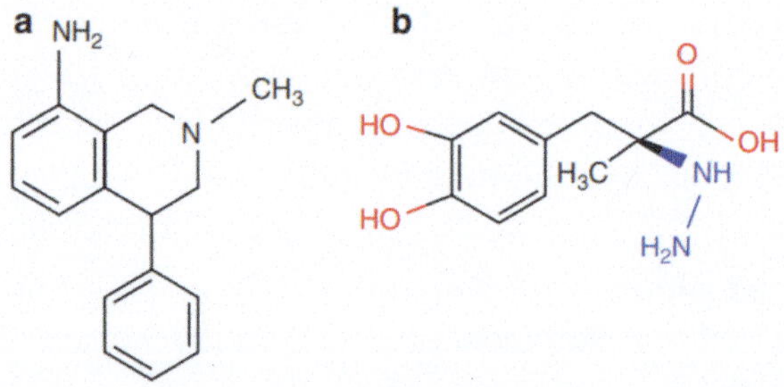

Figure 5.14. *The structural formula of (a) selegiline; and (b) carbidopa.*

Source: https://www.researchgate.net/figure/The-structures-of-selegiline-rasagiline-and-safinamide_fig2_321326859.

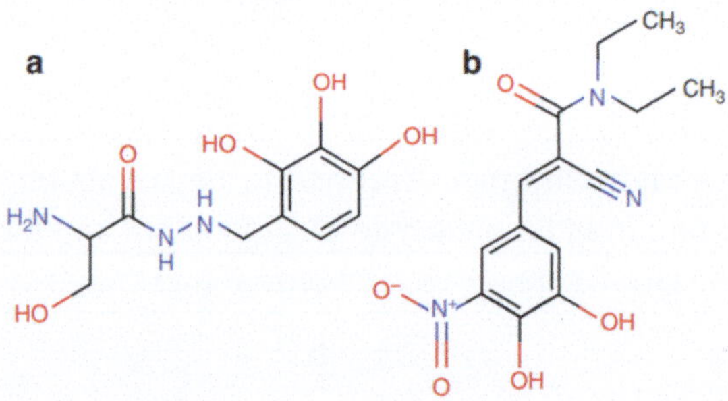

Figure 5.15. *Structural formula of (a) benserazide; and (b) entacapone.*

Source: https://www.researchgate.net/figure/Chemical-structures-of-a-levodopa-b-carbidopa-c-benserazide-hydrochloride-and-d_fig1_319137718.

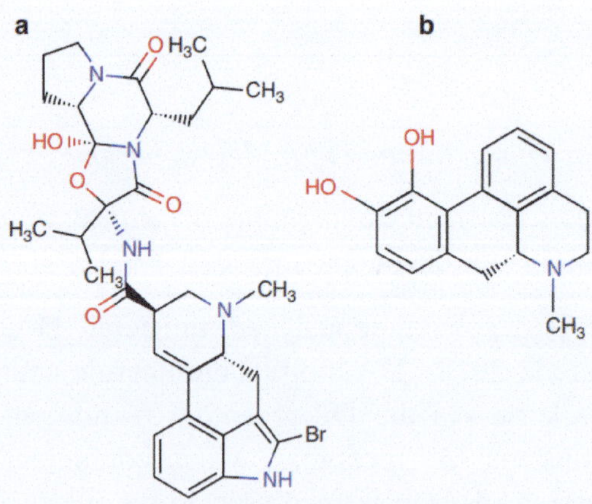

Figure 5.16. *The structural formula of (a) bromocriptine; and (b) apomorphine.*

Source: https://www.sciencedirect.com/topics/chemistry/ropinirole.

5.6.1.2. Alzheimer's Disease (AD)

AD is a progressive disease that causes emaciation and death of brain cells. It is the most common form of dementia, and it causes a decline in thinking, mental, and social abilities, interfering with a person's ability to function independently.

Degenerating neurons and connections in the frontal lobe and subcortical regions are signs of AD. The medications authorized for Alzheimer's treatment include cholinesterase inhibitors and memantine.

Treatment of Alzheimer's disease includes the following:

- Neurotransmitter manipulation that has been lost or decreased due to neurodegeneration;
- Factors that cause amyloid development are reduced and reversed.

In order to increase ACh activity, the following methods can be used: choline precursor administration to promote synthesis, cholinesterase inhibition to prevent degradation, and muscarinic and nicotinate receptor agonist to reproduce action. The illness is characterized by a diminution in the function of the CN. By decreasing the rate of breakdown of ACh, cholinergic acetylcholinesterase inhibitors are able to raise the content of ACh in the CNS while also preventing ACh loss affected by the deterioration of CNs.

Aminoacridine, velnacrine, and tacrine are cholinesterase inhibitors that are centrally active and reversible in action (Crismon, 1994). Donepezil is likewise a piperidine derivative with reversible acetylcholinesterase inhibitor action (Knowles, 2006), which was developed expressly for this purpose. An acetylcholinesterase inhibitor, also known as pseudo-irreversible rivastigmine, has the same function as ACh by attaching to the enzyme AChe (acetylcholinesterases) and forming a carbamylated complicated, which prevents ACh from binding (Birks and Grimley, 2015).

Acetate acetylcholinesterase inhibitor metrifonate is a permanent organophosphate acetylcholinesterase inhibitor (López and Schneider, 2006). ACEtylcholinesterases are enzymes that degrade ACh at the synaptic level, and galantamine has been demonstrated to improve the quantity of ACh at the synaptic level (Olin and Schneider, 2002).

5.6.2. Muscarinic Agonists

Amyloid plaques diminish the ability of muscarinic receptors to transfer signals, resulting in a reduction in cholinergic activity. Muscarinic receptors, on the other hand, are persistent and have therefore become an effective therapeutic target for improving cognitive performance in patients with illnesses. Drugs like bethanechol, carbachol, and methacholine are cholinesterase resistant but chemically similar to Ach.

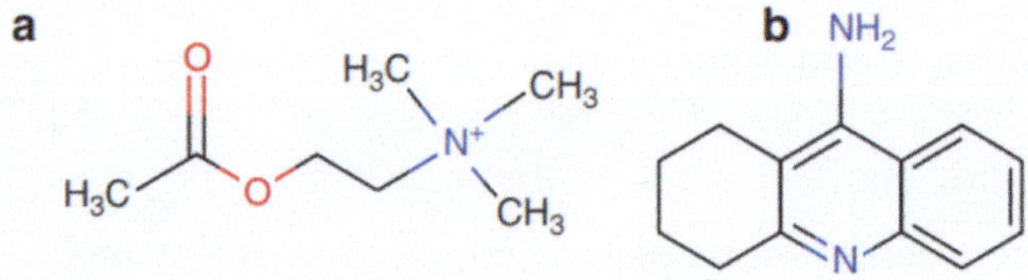

Figure 5.17. *The structural formula of (a) acetylcholine; and (b) tacrine.*

Source: https://www.sciencedirect.com/science/article/abs/pii/S004520681931569X.

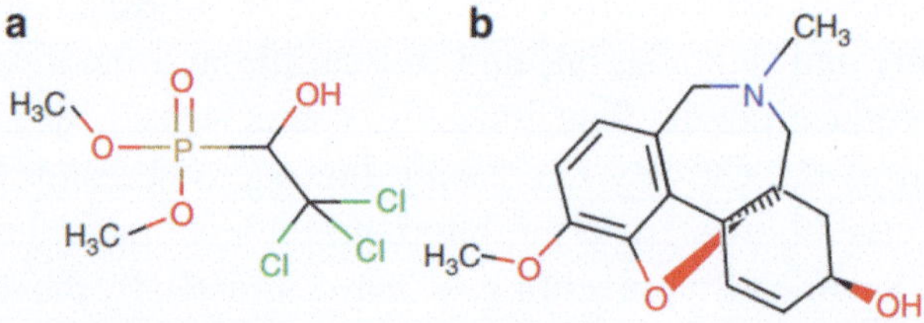

Figure 5.18. *Structural formula of (a) galantamine; and (b) metrifonate.*

Source: https://www.researchgate.net/figure/Traditional-cholinesterase-inhibitors-The-molecular-structures-of-a-physostigmine-B_fig2_333706502.

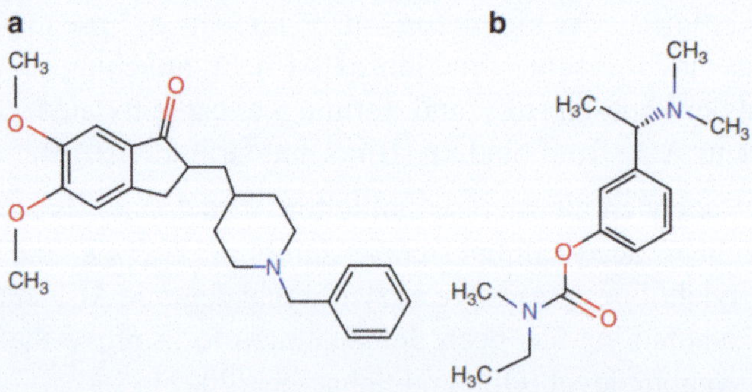

Figure 5.19. *Structural formula of (a) donepezil; and (b) rivastigmine.*

Source: https://www.researchgate.net/figure/Fig-1-Chemical-structures-of-Rivastigmine-Galantamine-HuperzineA-Donepezil-and_fig1_233588102.

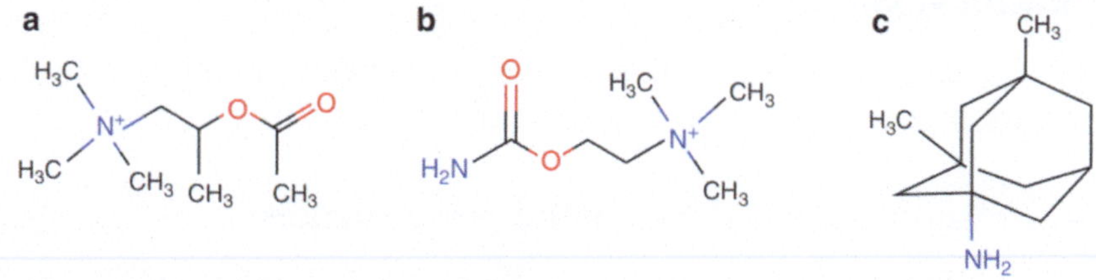

Figure 5.20. *Structural formula of (a) methacholine; (b) carbachol; and (c) memantine.*

Source: https://www.slideshare.net/coolboy101pk/cholinergic-agonists.

Nicotine also increases ACh release by stimulating presynaptic receptors on cholinergic nerve terminals. Because of increased oxidative stress related to amyloid-beta peptides, glutamatergic mechanisms play a vital role in fast synaptic transmission, learning, and memory operations, which are markedly disturbed in AD (Butterfield et al., 2003). Figures 5.17–5.20 show the structural formulas of medications used to treat AD.

5.6.2.1. Epilepsy

This is an episodic nervous system condition caused by a cluster of neurons that fire in a synchronized and continuous manner. Seizures can be avoided by lowering excitement or increasing neural inhibition. This is accomplished by:

- By blocking or activating GABA receptors, you can increase the quantity of gamma-aminobutyric acid (GABA);
- Blocking calcium or sodium channels that are excitatory voltage-gated;
- The antagonistic chloride channel is opened;
- Glutamate output or its N-methyl-D-aspartate (NMDA) receptors are reduced.

Phenytoin has no effect on the membrane's resting potential, but it does diminish the discharge induced by reversing the neuron while keeping the initial action potential intact.

It causes action by stopping the voltage-gated sodium network; that is, it makes the voltage-gated sodium channel inactive, keeping it inactive and unresponsive.

5.6.3. Ethosuximide

T-type calcium channels in thalamic neurons, which cause epileptic spikes and waves, appeared to be repressed (Glauser et al., 2010).

5.6.4. Benzodiazepines (BZD) and Barbiturates

These medicines exert their activity by activating the chloride ion channel through the action of GABA-A-type agonists. Phenobarbitone works directly on the Cl channel to prolong its opening time, whereas BZD work by altering the receptor's allosteric properties to increase the frequency at which the channel opens (Vining, 1986; Bianchi et al., 2009).

5.6.5. Valproic Acid/Sodium Valproate

It has been observed that the enzyme gamma-aminobutyric acid transaminase (GABA transaminase) is inhibited, resulting in increased inhibition and concentrations of GABA. Anti-seizure effects, on the other hand, manifest much more rapidly than this effect.

5.6.6. Lamotrigine

Lamotrigine is an antiepileptic medication of the next generation. Experiments have demonstrated that glutamate release and hence its stimulating effect are diminished (Gangurde et al., 2019).

5.6.7. Vigabatrin (Gamma-Vinyl GABA)

Because it has a molecular structure similar to GABA, it interacts with GABA transaminase and has an irreversible effect on it, resulting in an antiepileptic action. It has also been demonstrated to reduce glutamate levels.

5.6.8. Tiagabine

It is a lipophilic nipecotic acid derivative that prevents GABA from being absorbed by neurons and glia, resulting in improved extracellular GABA, and extended post-excitatory neuronal hyperpolarization.

5.6.9. Gabapentin

Gabapentin was formerly thought to be a cyclohexane analog of GABA, a receptor agonist, but it has been demonstrated to boost GABA levels via blocking GABA transferase and amplifying glutamate decarboxylase (GAD).

Figures 5.21 and 5.22 illustrate the structural formulas of medications used to treat epilepsy.

5.6.9.1. Depression

An alternative theory has been presented as a source of depression. A functional shortage of monoamine NTs such as 5-hydroxytryptamine (5-HT) A alternative theory has been presented as a source of depression.

A functional shortage of monoamine NTs such as 5-HT and adrenaline might induce it, according to the oldest and most commonly recognized explanation. Figure 5.23 shows the chemical structure of adrenaline.

Depression is also linked to a drop in brain-derived neurotropic aspect levels or dysfunction of its receptor TrkB. Noradrenergic and 5-HT, which govern the release of chemicals from this cell, are also input to hypothalamic neurons that control pituitary activity.

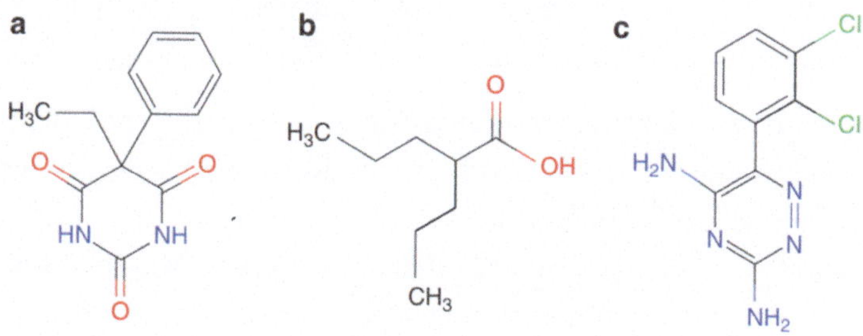

Figure 5.21. *Structural formula of (a) phenobarbitone; (b) valproic acid; and (c) lamotrigine.*

Source: https://www.mdpi.com/1424-8247/3/6/1909/htm.

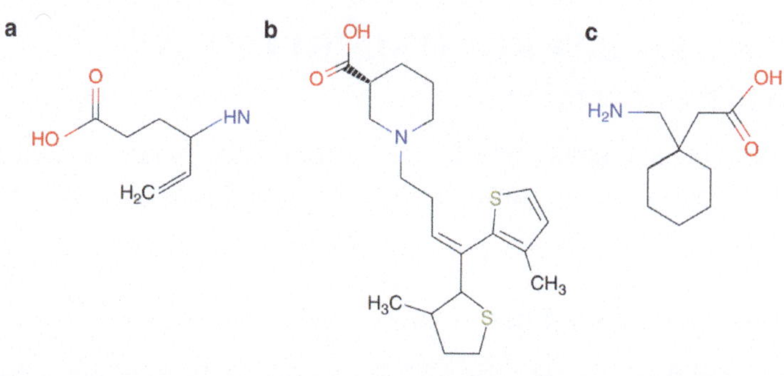

Figure 5.22. *Structural formula of (a) vigabatrin; (b) tiagabine; and (c) gabapentin.*

Source: https://www.researchgate.net/figure/Chemical-structures-of-the-anti-epileptic-drugs-discussed-in-this-review_fig1_45272592.

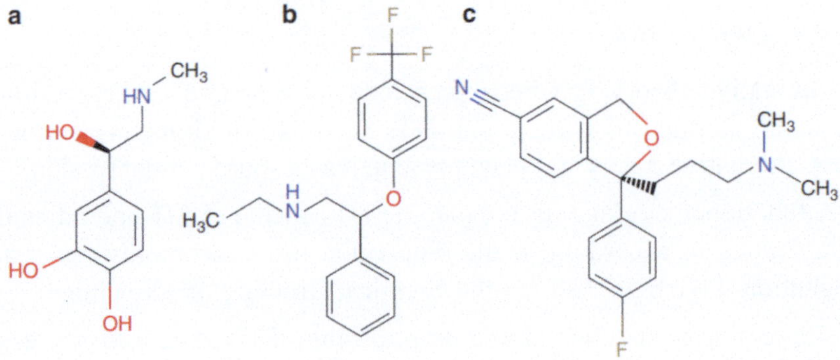

Figure 5.23. *Structural formula of (a) adrenaline; (b) fluoxetine; and (c) escitalopram.*

Source: https://www.researchgate.net/figure/Chemical-structures-of-SSRIs-and-SNRI-antidepressant-drugs_fig2_235903967.

Corticotrophin-releasing hormones are secreted by hypothalamic cells, which encourage pituitary cells to produce adrenocorticotropic hormones, which result in cortex secretion.

Antidepressants are a type of drug that is frequently used for mental health and depression. Tricyclic antidepressants (TCAs), atypical antidepressants, selective serotonin reuptake inhibitors (SSRIs), and monoamine oxidase inhibitors (MAOIs) are among them. Thus, antidepressants work primarily by raising the concentration of the monoamines 5-HT and noradrenaline in the synaptic cleft and their availability for binding (Delgado, 2000).

5.7. SELECTIVE SEROTONIN REUPTAKE INHIBITORS (SSRIS)

Antidepressants such as SSRIs are often administered. Citalopram, paroxetine, fluoxetine, sertraline, and escitalopram are examples of antidepressants.

5.7.1. Fluoxetine and Paroxetine

There are several conditions for which it is utilized, including depression, panic disorder, bulimia, obsessive-compulsive disorder (OCD), post-traumatic disorder, and premenstrual dysphoric disorder.

For clinical depression or generalized anxiety disorder, escitalopram is prescribed, and it has been demonstrated to be more efficient than citalopram.

It is frequently given to treat the major depressive disorder, OCD, panic disorder, premenstrual dysphoric disorder, post-traumatic stress disorder, and social anxiety disorders, among other conditions. It has been discovered that type escitalopram is more effective than all other SSRIs. Figure 5.23 depicts the structural formulas for the drugs adrenaline, fluoxetine, and escitalopram (Etain and Bonnet, 2001).

5.8. SEROTONIN-NOREPINEPHRINE REUPTAKE INHIBITOR (SNRI)

SNRIs (serotonin and norepinephrine reuptake inhibitors) are a class of medications that are effective in treating depression. Long-term (chronic) pain and anxiety problems, particularly nerve pain, are also treated with these. SNRIs inhibit the reabsorption (reuptake) of the NT's epinephrine and serotonin in the brain. Duloxetine and venlafaxine are two examples. Figure 5.24 shows the structural formula for this.

5.8.1. Duloxetine and Venlafaxine

They are employed to treat the main depressive disorder, neuropathic pain, fibromyalgia, and anxiety, with venlafaxine being utilized to give social phobia especially.

5.8.2. Mono-Amino Oxidase Inhibitor

This works by blocking the enzymes MAO A and MAO B, which cause monoamine neurochemicals like isocarboxazid to degrade. It has been used to treat depression, anxiety, Parkinson's disease, and dementia.

5.8.3. Schizophrenia

The individual and the planet are gradually disintegrating. Antipsychotics, also known as neuroleptics, are drugs that are used to treat symptoms of psychosis or schizophrenia.

Antipsychotics, commonly known as neuroleptic medications, are classed as stated in Table 5.1 to alleviate nervous tension. Figure 5.25 depicts the structural formula of medicines.

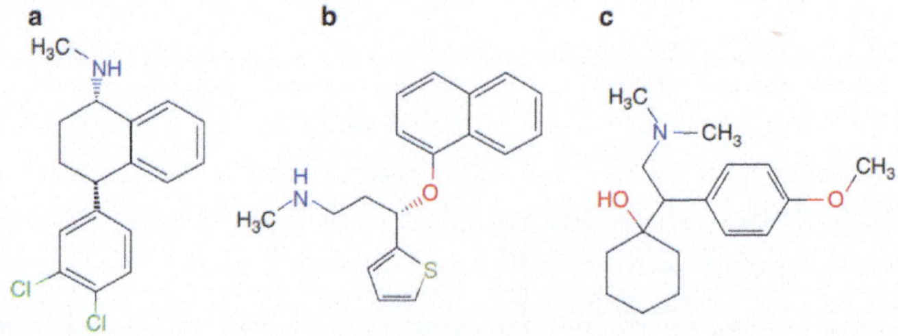

Figure 5.24. *The structural formula of (a) sertraline; (b) duloxetine; and (c) venlafaxine.*

Source: https://www.researchgate.net/figure/Chemical-structures-of-SSRIs-and-SNRI-antidepressant-drugs_fig2_235903967.

Table 5.1. List of Neuroleptic Agents

Atypical Antipsychotics	Typical Antipsychotics
Zotepine	Chlormazapine
Ziprasidone	Flupentixol
Clozapine	Haloperidol
Risperidone	Thiothiazine
Lurasidone	Trifluoperazine
Quetiapine	Perphenazine

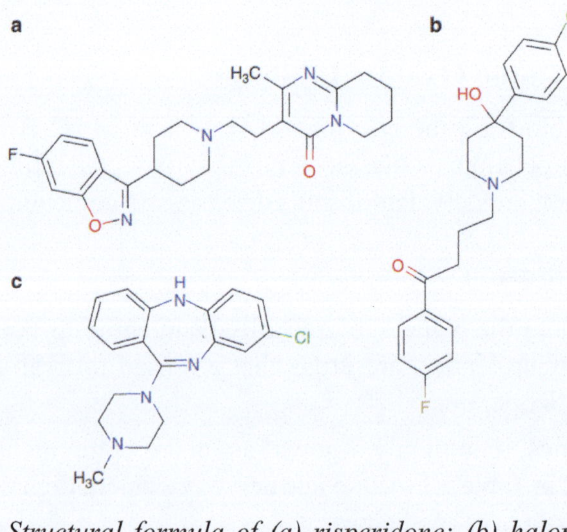

Figure 5.25. *Structural formula of (a) risperidone; (b) haloperidol; and (c) clozapine.*

Source: https://www.researchgate.net/figure/Scheme-1-Haloperidol-1-clozapine-2-risperidone-3-olanzapine-4-quetiapine-5_fig1_343522818.

They work by disrupting the DA system. Acute neuroleptics accelerate DA neuron discharge and so DA release. Because of the DA opponents, this is the case:

- Block the end autoreceptor's inhibitory impact on released DA;
- With comparable inhibitory receptors, block the activity of DA on the DA neuron;
- Block postsynaptic DA receptors on neurons that are inhibited by released DA, resulting in positive feedback to DA neurons.

As a result, D2 receptor binding is responsible for the majority of the effects. By inhibiting MAO enzymes, drugs such as haloperidol and chlormazapine enhance monoamine content in the brain, and as a result, they have been implicated in the treatment of schizophrenia. Drugs such as risperidone work by inhibiting the DA D2 receptor postsynaptically without changing DA levels in the brain or body.

5.8.4. Hypnosis and Anxiety

Anxiety disorders are considered to be one of the more traditional sorts of ailments. Anxiety is primarily caused by the release of neurochemicals, for example, norepinephrine, GABA, and serotonin in the brain. Since the beginning of time, alcohol has been the most extensively utilized antipsychotic agent. Barbiturates, such as phenobarbitone and meprobamate, eventually took the role of this therapy.

Barbiturates connect non-competitively to a distinct receptor domain, which may be coupled completely to the chloride channel in a number of different ways. While exhibiting an allosteric relationship with GABA, it also has the additional effect of increasing Cl conductivity by increasing the duration of the channel opening.

Diazepam establishes a connection in the brain to a specific location. It has been determined that the binding site is a GABA-A-type receptor element that is concerned with the opening of the Cl-channel by solubilized receptor studies. Increased drug binding is caused by allosteric communication between the GABA-A-type receptor and the diazepam receptor. This is assumed to be the result of an interaction between both the GABA recognition site on a -subunit GABA receptor and the benzodiazepine recognition site on an a-subunit benzodiazepine receptor (Cipriani et al., 2018).

The overall function of BZD is to enhance the rise in GABA-induced Cl ion conductance, hence increasing the inhibitory activities of the drug. This is accomplished by increasing the possibility and, as a result, the frequency of Cl ion channel openings.

The imidazopyridine zolpidem (insomnia medication), which was formerly thought to be a ligand for the BZ1 receptor, has now been identified as a high-efficiency ligand for the GABA-A receptor a1 subunit, which is a novel finding. Figure 5.26 depicts the chemical structure of some commonly used medicines in the treatment of anxiety and hypnosis (Schneier, 1990).

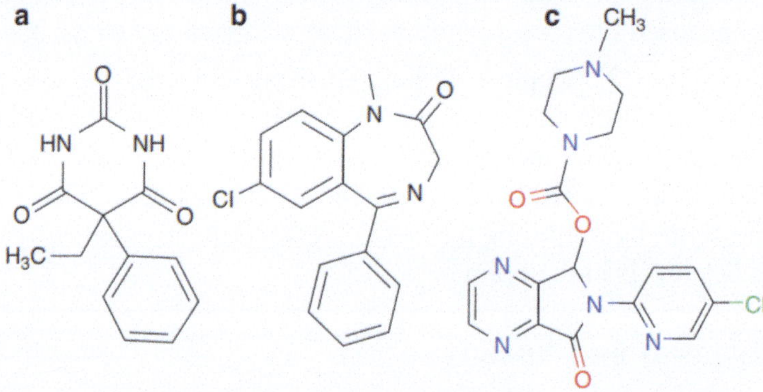

Figure 5.26. *Structural formula of (a) phenobarbitone; (b) diazepam; and (c) zolpidem.*

Source: https://www.researchgate.net/figure/Structure-of-diazepam-bretazenil-imidazenil-and-other-imidazo-benzodiazepine_fig1_41942953.

5.9. SUMMARY

Among the most thoroughly researched topics of research, the study of psychopharmaceuticals, neuropeptides, and their receptors is one that has a tremendous amount of research data accessible. Learning about the neuropeptide signaling pathways and the location of neuropeptide receptors in the brain nervous system, particularly in the region of significant variable tasks, provides a solid foundation for apparent notice in ligands of neurochemical receptors, such as agonists, antagonists, and modulators of neurochemical receptors, among other things. Despite the large number of medical trials that have been conducted, the cumulative number of genetic research that has revealed links between neuropeptide and disease signaling reinforces the belief that many neuropeptide receptors and ligands for

NTs will be developed as drugs in the future. A greater understanding of neurochemical activities and the signaling pathways that regulate them will pave the way for considerable future discovery and development.

REFERENCES

1. Agapitov, A. V., & Haynes, W. G., (2002). Role of endothelin in cardiovascular disease. *J Renin- Angiotensin- Aldosterone Syst., 3*(1), 1–15.
2. Alescio-Lautier, B., & Soumireu-Mourat, B., (1998). Role of vasopressin in learning and memory in the hippocampus. *Prog. Brain Res., 119*, 501–521.
3. Bianchi, M. T., Botzolakis, E. J., Lagrange, A. H., & Macdonald, R. L., (2009). Benzodiazepine modulation of $GABA_A$ receptor opening frequency depends on activation context: A patch-clamp and simulation study. *Epilepsy Res., 85*(2, 3), 212–220.
4. Birks, J. S., & Grimley, E. J., (2015). Rivastigmine for Alzheimer's disease. *Cochrane Database Syst. Rev., 10*(4), CD001191.
5. Brighton, P. J., Szekeres, P. G., & Willars, G. B., (2004). Neuromedin U and its receptors: Structure, function, and physiological roles. *Pharmacol. Rev., 56*(2), 231–248.
6. Burbach, J. P., (2011). What are neuropeptides? *Methods Mol. Biol., 789*, 1–36.
7. Butterfield, D. A., Boyd-Kimball, D., & Castegna, A., (2003). Proteomics in Alzheimer's disease: Insights into potential mechanisms of neurodegeneration. *J. Neurochem., 86*(6), 1313–1327.
8. Calo' G., Guerrini, R., Rizzi, A., Salvadori, S., & Regoli, D., (2000). Pharmacology of nociceptin and its receptor: A novel therapeutic target. *Br. J. Pharmacol., 129*(7), 1261–1283.
9. Cipriani, A., Furukawa, T. A., Salanti, G., Chaimani, A., Atkinson, L. Z., Ogawa, Y., Leucht, S., et al., (2018). Comparative efficacy and acceptability of 21 antidepressant drugs for the acute treatment of adults with major depressive disorder: A systematic review and network meta-analysis. *Lancet, 391*(10128), 1357–1366.
10. Clement-Jones, V., Lowry, P. J., Rees, L. H., & Besser, G. M., (1980). Met-enkephalin circulates in human plasma. *Nature, 283,* 295–297.
11. Córdoba-Chacón, J., Gahete, M. D., Culler, M. D., Castaño, J. P., Kineman, R. D., & Luque, R. M., (2012). Somatostatin dramatically stimulates growth hormone release from primate somatotrophs acting at low doses via somatostatin receptor 5 and cyclic AMP. *J Neuroendocrinol., 24*(3), 453–463.
12. Crismon, M. L., (1994). Tacrine: First drug approved for Alzheimer's disease. *Ann. Pharmacother., 28*(6), 744–751.

13. De Girolamo, P., & Dieguez, C., (2014). Neuropeptides and control of food intake. *Int. J. Endocrinol. 2014*, 910912.
14. De Mota, N., Lenkei, Z., & Llorens-Cortès, C., (2000). Cloning, pharmacological characterization and brain distribution of the rat apelin receptor. *Neuroendocrinology, 72*(6), 400–407.
15. Delgado, P. L., (2000). Depression: The case for a monoamine deficiency. *J. Clin. Psychiatry, 61*(6), 7–11.
16. Diaz-Del, C. M., Woldbye, D. P. D., & Heegaard, A. M., (2018). Neuropeptide Y and its involvement in chronic pain. *Neuroscience, 387*, 162–169.
17. Etain, B., & Bonnet-Perrin, E., (2001). Value of fluoxetine in obsessive-compulsive disorder in the adult: Review of the literature. *L'Encéphale, 27*(3), 280–289.
18. Evans, H., Baumgartner, M., Shine, J., & Herzog, H., (1993). Genomic organization and localization of the gene encoding human preprogalanin. *Genomics, 18*(3), 473–477.
19. Gangurde, P. K., Ajitkumar, B. N., & Kumar, L., (2019). Lamotrigine lipid nanoparticles for effective treatment of epilepsy: A focus on brain targeting via nasal route. *J. Pharm. Innov., 14*, 91–111.
20. Gavioli, E. C., Vaughan, C. W., Marzola, G., Guerrini, R., Mitchell, V. A., Zucchini, S., De Lima, T. C., et al., (2004). Antidepressant-like effects of the nociceptin/orphanin FQ receptor antagonist UFP-101: New evidence from rats and mice. *Naunyn Schmiedeberg's Arch Pharmacol., 369*(6), 547–553.
21. Glauser, T. A., Cnaan, A., Shinnar, S., Hirtz, D. G., Dlugos, D., Masur, D., Clark, P. O., et al., (2010). Ethosuximide, valproic acid, and lamotrigine in childhood absence epilepsy. *N Engl. J. Med., 362*, 790–799.
22. Hökfelt, T., Broberger, C., Xu, Z. Q., Sergeyev, V., Ubink, R., & Diez, M., (2000). Neuropeptides—An overview. *Neuropharmacology, 39*(8), 1337–1156.
23. Inutsuka, A., & Yamanaka, A., (2013). The physiological role of orexin/hypocretin neurons in the regulation of sleep/wakefulness and neuroendocrine functions. *Front Endocrinol., 4*, 18.
24. Jankovic, J., (2008). Parkinson's disease: Clinical features and diagnosis. *J. Neurol. Neurosurg. Psychiatry, 79*(4), 368–376.
25. Jenck, F., Moreau, J. L., Martin, J. R., Kilpatrick, G. J., Reinscheid, R. K., Monsma, F. J., Nothacker, H. P., et al., (1997). Q acts as an anxiolytic to attenuate behavioral responses to stress. *Proc. Natl. Acad Sci. U S A, 94*(26), 14854–14858.

26. Katzenschlager, R., Poewe, W., Rascol, O., Trenkwalder, C., Deuschl, G., Chaudhuri, K. R., Henriksen, T., et al., (2018). Apomorphine subcutaneous infusion in patients with Parkinson's disease with persistent motor fluctuations (TOLEDO), a multicentre, double-blind, randomized, placebo-controlled trial. *Lancet Neurol., 17*(9), 749–759.

27. Kline, N. S., (1959). Psychopharmaceuticals: Effects and side effects. *Bull. Org. Mond. Sante, 21*, 397–410.

28. Knowles, J., (2006). Donepezil in Alzheimer's disease: An evidence-based review of its impact on clinical and economic outcomes. *Core Evid., 1*(3), 195–219.

29. Kuhn, H., Hubbard, A., Kannengieser, J., & Hackworth, C., (2006). The role of endothelin in stroke. *GUJHS, 3*, 1. https://blogs.commons.georgetown.edu/journal-of-health-sciences/issues-2/previous-volumes/vol-3-no-1-march-2006/the-role-of-endothelin-in-stroke/ (accessed on 23 March 2022).

30. Lee, D. A., Andreev, A., Truong, T. V., Chen, A., Hill, A. J., Oikonomou, G., Pham, U., et al., (2017). Genetic and neuronal regulation of sleep by neuropeptide VF. *eLife, 6*, e25727.

31. Lin, L. C., & Sibille, E., (2013). Reduced brain somatostatin in mood disorders: A common pathophysiological substrate and drug target? *Front Pharmacol., 4*, 12. Article No.110.

32. López-Arrieta, J., & Schneider, L., (2006). Metrifonate for Alzheimer's disease. *Cochrane Database Syst. Rev.* (2), CD003155. https://doi.org/10.1002/14651858.CD003155.pub3.

33. Lundström, L., Elmquist, A., Bartfai, T., & Langel, U., (2005). Galanin and its receptors in neurological disorders. *Neuromolecular Med., 7*(1, 2), 157–180.

34. Mahoney, C. E., Cogswell, A., Koralnik, I. J., & Scammell, T. E., (2019). The neurobiological basis of narcolepsy. *Nat. Rev. Neurosci., 20*(2), 83–93.

35. Maixnerová, J., Špolcová, A., Pýchová, M., Blechová, M., Elbert, T., Rezáčová, M., Zelezná, B., & Maletínská, L., (2011). Characterization of prolactin-releasing peptide: Binding, signaling and hormone secretion in rodent pituitary cell lines endogenously expressing its receptor. *Peptides, 32*(4), 811–817.

36. Martinez, V. G., O'Driscoll, L., & Neuromedin, U., (2014). A multifunctional neuropeptide with pleiotropic roles. *Clin. Chem., 61*(3), 471–482.

37. Melas, P. A., Guban, P., Rahman, M. S., Lavebratt, C., & Forsell, Y., (2018). Neuropeptide Y, stressful life events and personality trait

conscientiousness: Preliminary associations from a Swedish longitudinal study. *Psychiatry Res., 263*, 48–53.

38. Meunier, J. C., Mollereau, C., Toll, L., Suaudeau, C., Moisand, C., Alvinerie, P., Butour, J. L., et al., (1995). Isolation and structure of the endogenous agonist of opioid receptor-like ORL1 receptor. *Nature, 377*(6549), 532–535.
39. Mitsukawa, K., Lu, X., & Bartfai, T., (2008). Galanin, galanin receptors and drug targets. *Cell Mol. Life Sci., 65*(12), 1796–1805.
40. Niermann, H., Amiry-Moghaddam, M., Holthoff, K., Witte, O. W., & Ottersen, O. P., (2001). A novel role of vasopressin in the brain: Modulation of activity-dependent water flux in the neocortex. *J. Neurosci., 21*(9), 3045–3051.
41. O'Carroll, A. M., Selby, T. L., Palkovits, M., & Lolait, S. J., (2000). Distribution of mRNA encoding B78/apj, the rat homolog of the human APJ receptor, and its endogenous ligand apelin in brain and peripheral tissues. *Biochim. Biophys. Acta, 1492*(1), 72–80.
42. Ohki-Hamazaki, H., (2000). Neuromedin B. *Prog. Neurobiol., 62*(3), 297–312.
43. Okuda-Ashitaka, E., Minami, T., Tachibana, S., Yoshihara, Y., Nishiuchi, Y., Kimura, T., & Ito, S., (1998). Nocistatin, a peptide that blocks nociceptin action in pain transmission. *Nature, 392*(6673), 286–289.
44. Olin, J., & Schneider, L., (2002). Galantamine for Alzheimer's disease. *Cochrane Database Syst. Rev., 3*, CD001747.
45. Russo, A. F., (2017). Overview of neuropeptides: Awakening the senses? *Headache, 57*(Suppl. 2), 37–46.
46. Sakurai, T., (2014). The role of orexin in motivated behaviors. *Nat. Rev. Neurosci., 15*(11), 719–731.
47. Schinelli, S., (2006). Pharmacology and physiopathology of the brain endothelin system: An overview. *Curr. Med. Chem., 13*(6), 627–638.
48. Schneier, F. R., Liebowitz, M. R., Davies, S. O., Fairbanks, J., Hollander, E., Campeas, R., & Klein, D. F., (1990). Fluoxetine in panic disorder. *J. Clin. Psychopharmacol., 10*(2), 119–121.
49. Shuler, M. L., & Hickman, J. J., (2014). Toward *in vitro* models of brain structure and function. *Proc. Natl. Acad Sci. U S A, 111*(38), 13682, 13683.
50. Singh, C., Rihel, J., & Prober, D. A., (2017). Neuropeptide Y regulates sleep by modulating noradrenergic signaling. *Curr. Biol., 27*(24), 3796–3811.e5.
51. St-Gelais, F., Jomphe, C., & Trudeau, L. E., (2006). The role of neurotensin

in central nervous system pathophysiology: What is the evidence? *J. Psychiatry Neurosci., 31*(4), 229–245.

52. Stockbrügger, R. W., (1989). Antimuscarinic drugs. *Methods Find Exp. Clin. Pharmacol., 11*(1 Suppl), 79–86.
53. Subirán, N., Candenas, L., Pinto, F. M., Cejudo-Roman, A., Agirregoitia, E., & Irazusta, J., (2012). Autocrine regulation of human sperm motility by the met-enkephalin opioid peptide. *Fertil. Steril. 98*(3), 617–625.e3.
54. Sunter, D., Hewson, A. K., & Dickson, S. L., (2003). Intracerebroventricular injection of apelin-13 reduces food intake in the rat. *Neurosci. Lett., 353*(1), 1–4.
55. Tachibana, T., & Sakamoto, T., (2014). Functions of two distinct "prolactin-releasing peptides" evolved from a common ancestral gene. *Front Endocrinol., 10*(5), 170.
56. Taheri, S., Murphy, K., Cohen, M., Sujkovic, E., Kennedy, A., Dhillo, W., Dakin, C., et al., (2002). The effects of centrally administered apelin-13 on food intake, water intake and pituitary hormone release in rats. *Biochem. Biophys. Res. Commun., 291*(5), 1208–1212.
57. Tatemoto, K., Hosoya, M., Habata, Y., Fujii, R., Kakegawa, T., Zou, M. X., Kawamata, Y., et al., (1998). Isolation and characterization of a novel endogenous peptide ligand for the human APJ receptor. *Biochem. Biophys. Res. Commun., 251*(2), 471–476.
58. Toll, L., Bruchas, M. R., Calo,' G., Cox, B. M., & Zaveri, N. T., (2016). Nociceptin/orphanin FQ receptor structure, signaling, ligands, functions, and interactions with opioid systems. *Pharmacol. Rev., 68*(2), 419–457.
59. Tost, H., Alam, T., & Lindenberg, A. M., (2010). Dopamine and psychosis: Theory, pathomechanisms and intermediate phenotypes. *Neurosci. Biobehav. Rev., 34*(5), 689–700.
60. Van, D. P. A. N., (2012). Neuropeptide transmission in brain circuits. Neuron 76(1), 98–115.
61. Vining, E. P. G., (1986). Use of barbiturates and benzodiazepines in the treatment of epilepsy. *Neurol. Clin., 4*(3), 617–632.
62. Wang, C., Wang, Q., Ji, B., Pan, Y., Xu, C., Cheng, B., Bai, B., & Chen, J., (2018). The orexin/receptor system: Molecular mechanism and therapeutic potential for neurological diseases. *Front Mol. Neurosci., 11*, 220.
63. Wang, Y., Wang, M., Yin, S., Jang, R., Wang, J., Xue, Z., & Xu, T., (2015). *NeuroPep: A Comprehensive Resource of Neuropeptides.* Database (Oxford). Bav038. https://doi.org/10.1093/database/bav038. http://isyslab.info/NeuroPep/.

64. Watkins, L. E., Han, S., Krystal, J. H., Southwick, S. M., Gelernter, J., & Pietrzak, R. H., (2017). Association between functional polymorphism in neuropeptide Y gene promoter rs16147 and resilience to traumatic stress in us military veterans. *J. Clin. Psychiatry, 78*(8), e1058–e1059.
65. Wysocka, M. B., Pietraszek-Gremplewicz, K., & Nowak, D., (2018). The role of apelin in cardiovascular diseases, obesity and cancer. *Front Physiol., 23*(9), 557.
66. Yamanaka, A., Tabuchi, S., Tsunematsu, T., Fukazawa, Y., & Tominaga, M., (2010). Orexin directly excites orexin neurons through orexin two receptor. *J. Neurosci., 30*(38), 12642–12652.
67. Zaveri, N. T., (2011). The nociceptin/orphanin FQ receptor (NOP). as a target for drug abuse medications. *Curr. Top. Med. Chem., 11*(9), 1151–1156.

CHAPTER 6

Neurochemical Communication in Endocannabinoids

Contents

6.1. INTRODUCTION

The brain has a vast array of molecules that enable neurons to interact with one another. A few of these molecules, like glutamate and GABA, are recognized to serve as neurotransmitters (NTs) for years, while the others have recently been discovered to play a role in brain signaling. Cannabinoids are a kind of cannabinoids. Although people have known about the impact of cannabinoids like dynamic elements in marijuana on the psyche of people and behavior for years, if not millennia, their occurrence and manufacturing in the human brain had only been discovered 30 years before. Their practical involvement in neuronal routes of the brain is just lately discovered. Endogenous cannabinoids, often known as endocannabinoids (eCBs), are endogenously manufactured signaling molecules. The function of these NT systems in the nervous system, as well as their interactions with other NT systems, would be discussed. The capacity of endocannabinoids to operate as retrograde messengers in neuronal circuits is a distinguishing property of their signaling. This intercellular communication passageway is illustrated in-depth in two cases, first from the hippocampus and the second from the primary olfactory bulb (Calignano et al., 2000).

Several Numerous characteristics highlight the necessity of comprehending the endocannabinoid system (Cota et al., 2003). The importance of endocannabinoids in typical behaviors, such as the reception of pain and eating is becoming clearer. Cannabinoids' therapeutic potential has gotten a lot of attention in the last some years (Iversen and Chapman, 2002; Soria-Gómez et al., 2014). Endocannabinoids are involved in neuroprotection in opposition to sensitive excitotoxicity as well as functional upturn following the injury of the brain (Marsicano et al., 2001, 2003). Endocannabinoids are a kind of cannabinoid that regulates human beings' function of the airway and may be used to medicate respiratory diseases (Katona, 2001). Cannabinoids are widely used as psychoactive substances for recreational purposes and may interact with further misuse of drugs. Such reality emphasizes the importance of learning more about the system of endocannabinoids and the neurobiological basis for its mood-changing abilities. Moreover, the system of endocannabinoids plays a critical part in memory and learning procedures, such as the elimination of painful memories (Valjent et al., 2002; Marsicano et al., 2002).

6.2. THE ENDOCANNABINOID SYSTEM

Endocannabinoids are tiny lipids that control a variety of elements of the function of the brain, like synaptic transmission and various kinds of long-term and short-term plasticity. They have an impact on synapse production and neurogenesis, as well as development and growth. Feeding and anxiety are two more biological activities influenced by endocannabinoids. The natural ligands/agonists of the extensively represented receptor of cannabinoids inside the brain are mainly two endocannabinoids, two-arachidonoylglycerol (2-AG) and N-arachidonoyl ethanol-amide (anandamide (AEA)) (Alger, 2001; Kano et al., 2009) (Figure 6.1).

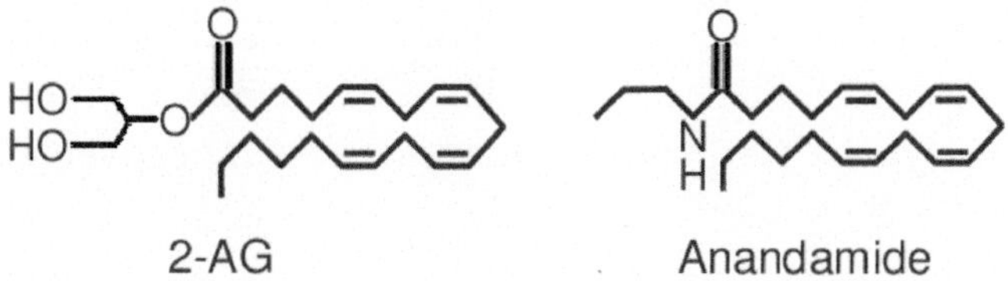

Figure 6.1. *The natural ligands/agonists of cannabinoid receptors, CB1Rs, in the brain are two endocannabinoids: N-arachidonoylethanol-amide and two-arachidonoylglycerol.*

Source: https://www.researchgate.net/figure/Two-endocannabinoids-2-arachidonoylglycerol-2-AG-and-N-arachidonoylethanol-amide_fig1_263662509.

The endocannabinoid system is made up of endogenous ligands produced from fatty acids, and G-protein attached receptors of cannabinoids. This system also comprises molecular machinery such as endocannabinoid precursors, lipidic NT synthesizers and degraders, and transporters (Alger, 2002). CB1 and CB2 receptors, two distinct cannabinoid receptors (CB1R), are cloned. They are 44% similar in amino acid sequence. The two CB1R have quite varied expression patterns in various areas of the body. CB1R is a very common G-protein attached receptor in the brain (Freund et al., 2003). Immune cells and peripheral organs are the primary sites of CB2R expression. The expression of CB2R is found in the cortex, cerebellar neurons, and brainstem, as well as in microglia (Alger & Kim, 2011).

In the brain, CB1R is abundant, particularly at the terminals of the presynaptic nerve. Cannabis-derived medications can activate them. THC, or 9-tetrahydrocannabinol, is the primary component in hashish and marijuana and may be used to activate CB1R artificially. When activated by their natural ligands, the receptors of cannabinoids are present in every ordinary brain, wherein they fulfill a variety of important brain activities. In the neurological system, CB1R are mostly $G_{i/o}$ protein-coupled group-1 receptors of cannabinoid (receptors of CB1, CB1Rs). Endocannabinoids, their ligands, have been made through the lipids of the membrane (Howlett,

2004). Endocannabinoids may disseminate across membranes and trigger receptors in a similar way as exogenously administered cannabinoids such as cannabis do. 2-AG and AEA had been found in the 1990s, but their functional involvement in neural transmission was unknown for many years. Endocannabinoids have been well-established as retrograde messengers that reduce both inhibitory and excitatory transmission since their discovery. In the olfactory bulb, cerebellum, hippocampus, amygdala, and neocortex, endocannabinoids regulate retrograde signals. Reabsorption of endocannabinoids into glia and neurons stops endocannabinoids from signaling. Fatty acid amide hydrolase (FAAH) and monoacylglycerol lipase (MAGL) hydrolyze 2-AG and AEA intracellularly, respectively (Matsuda et al., 1990; Munro et al., 1993).

6.3. NEW AND EXTRAORDINARY NEUROTRANSMITTERS (NTS)

Endocannabinoids vary as traditional NTs that have lipids, which are rapidly generated on request at the place of requirement by cell membrane elements instead of being stored. They have been discharged from throughout the cell when the cell is activated. Phospholipase activity produces arachidonic acid-containing messengers. The production of endocannabinoids is activated by a spike in the intracellular Ca levels within the neuron, as well as the activation of specific receptors of the type of G-protein-coupled, like metabotropic glutamate receptors (mGluR5). Endocannabinoids engage cannabinoid receptors on adjacent neurons, like presynaptic interneurons, after their non-synaptic, non-vesicular emission, and control presynaptic NT emission, for example, by closing certain ion pathways (Herkenham et al., 1990; Van et al., 2005).

6.4. DEPOLARIZATION-INDUCED SUPPRESSION OF INHIBITION (DSI)

The importance of the setup of endocannabinoids for neuronal signaling and the working of the brain is investigated. Endocannabinoids have a role in a novel kind of neural communication known as DSI (Figure 6.2). A brief increase in intracellular Ca concentration in the main neuron, such as hippocampal pyramidal cell, causes a momentary reduction in arriving inhibitory signals in the shape of GABA through other types of neurons. Such finding led to the theory of depolarization-induced suppression of inhibition

(DSI), an unidentified messenger should pass from the postsynaptic cell to the presynaptic GABA-discharging cell and shut down NT secretion in a certain way. Traditional chemical synaptic signaling among two neurons plays a role in the activation of a presynaptic neuron, which releases transmitters, and the succeeding postsynaptic neuron activation, for example, when inhibitory interneuron of GABAergic synaptically comes into contact using a pyramidal cell of glutamatergic in the hippocampus. The interneuron produces the hindering NT of GABA, which suppresses the pyramidal cell whenever it has been engaged. Whenever a pyramidal neuron is triggered, such as a direct injection current, the hindering input to the pyramidal cell is lowered during DSI. Endocannabinoids were discovered to operate as regressive molecules of signaling that facilitate transmission among the pyramidal cells of postsynaptic and inhibitory interneurons of presynaptic as well as obtain the depletion in GABA secretion, which was advancement in the knowledge of endocannabinoid signaling. Because endocannabinoids are lipids, they do not diffuse much in the brain's aqueous extracellular environment. DSI, on the other hand, has a temporary local impact that allows single neurons to temporarily separate from their neighbors to predetermine data (Núñez et al., 2004).

Alger (2001) used the phrase "Dies mirabilis" (beautiful day) to describe the news of this discovery. Four separate laboratories published their findings in three distinct publications in March 2001, concluding that endocannabinoids primarily serve as reverse messengers. Egorova and Elphick analyzed anatomical and pharmacological study results of the activities of cannabinoid receptor agonists and collected them using their information of CB1R localization and AEA degrading enzymes, FAAH, to conclude that endocannabinoids behave as reverse messengers. Ohno-Shosaku et al. (2001); and Wilson and Nicoll (2001) revealed that cannabinoid receptor activation resembled DSI, but cannabinoid receptor blockage inhibited DSI. On cerebellar excitatory synapses, Regehr and Kreitzer (2001) reported comparable phenomena; Depolarization induced Suppression of Excitation (DSE), regulated by reverse endocannabinoid activity. DSE and DSI are predicated on a presynaptic action, as evidenced by alterations in the paired-pulse ratio of NT secretion and a rise in Ca in the cells of postsynaptic.

Extracellular stimuli, like NTs, are transduced into intracellular signaling cascades by G-protein coupled receptors (GPCRs). Activation of certain GPCRs, such as DA, muscarinic M1/M3 receptors of acetylcholine (ACh), or metabotropic glutamate, promotes the emission of endocannabinoids for several minutes. Even though endocannabinoids are normally generated in a Ca-dependent way, no obvious increase in intracellular Ca is required in the mGluR-dependent and mAChR-dependent passageways (Herkenham et al., 1991; Matsuda et al., 1993).

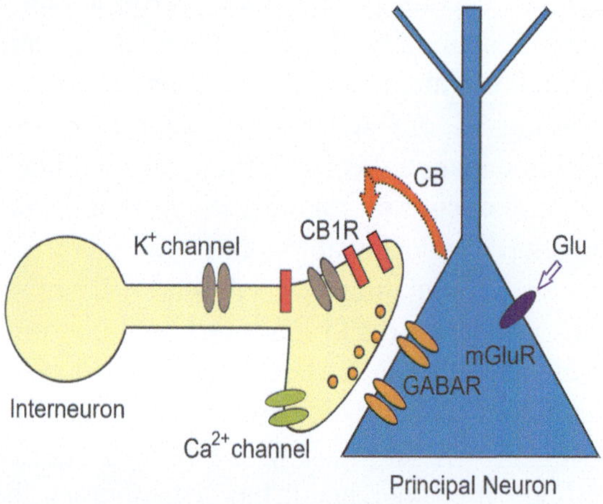

Figure 6.2. *DSI is a form for reverse signaling in the brain that enables researchers to measure actual endocannabinoid emission from main neurons in response to a momentary stoppage in GABA output.*

Note: Cannabinoids are synthesized and released when mGluRs on primary neurons are activated or when postsynaptic principal cells are depolarized (CBs). Cannabinoids combine to presynaptic cannabinoid receptors (CB1R) on the interneurons of GABAergic, reducing GABA emission from synaptic endpoints for a brief period. As a result, $GABA_A$ receptor-moderated currents of synaptic and inhibition of GABAergic in the main postsynaptic neurons are temporarily reduced.

Source: https://www.researchgate.net/figure/Depolarization-induced-Suppression-of-Inhibition-DSI-is-a-model-for-retrograde_fig2_263662509.

Though in the existence of higher intracellular calcium chelator concentrations, the secretion of endocannabinoids can begin, however, endocannabinoids can be responsive to the standard intracellular [Ca]. The stimulation of G-protein coupled receptor of postsynaptic cells increases the DSI, according to research conducted by the Alger lab and others. For example, glutamate acting on the group-I mGluRs straightforwardly creates endocannabinoids and increases the DSI. It is now well documented that the usage of endocannabinoids to convey or good instructions to target cells is associated with the stimulation of different GPCRs (Katona et al., 1999).

The DSI detection was a watershed moment in our knowledge which play a role in the functioning of the brain. Depolarization-induced suppression is a kind of temporary plasticity of synaptic that was first discovered in the hippocampus and cerebellum. Endocannabinoids are the molecules of reverse signaling that pass from main depolarized neurons to inhibitory interneurons of presynaptic to limit GABA secretion. DSI is a new regulatory pathway that causes a brief reduction of synaptic $GABA_A$ responses via retrograde endocannabinoid signaling from main neurons (Figure 6.3). Processed neurons adjust the strength of the synapses that are established onto them by retrograde signaling, and so modulate their synaptic excitableness in an activity-based way; this is significant in the data processing of neural systems. DSE is a retrograde signaling pathway in the cerebellum that lowers synaptic excitement by decreasing presynaptic glutamate release, comparable to DSI (Tsou et al., 1999).

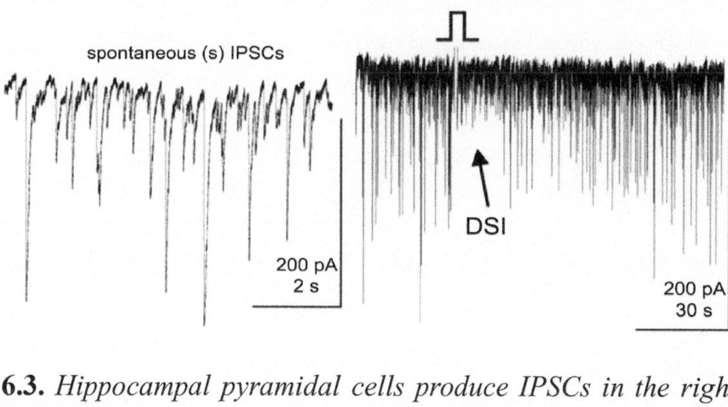

Figure 6.3. *Hippocampal pyramidal cells produce IPSCs in the right panel. Left panel: The pyramidal cell shows DSI, a transitory drop in IPSC activity caused by endocannabinoids acting on CB1R on presynaptic GABAergic interneurons, in response to a 1-s voltage pulse.*

Source: https://www.researchgate.net/publication/5960694_Involvement_of_Nitric_Oxide_in_Depolarization-Induced_Suppression_of_Inhibition_in_Hippocampal_Pyramidal_Cells_during_Activation_of_Cholinergic_Receptors.

6.5. HIPPOCAMPAL DSI

Endocannabinoids are lipids that, contrasting to traditional NTs, have been quickly generated from cell membrane components instead of being preserved. Postsynaptic somatodendritic regions, which are easily available to whole-cell patch electrodes, manufacture, and release them.

The impacts of such lipid signs are measured electrophysiologically as CB1R-based changes in traditional synaptic communication, providing a responsive method of bioassay in endocannabinoid activities and degrees. Ca^{2+}-dependent or comparatively Ca^{2+}-independent mechanisms may cause endocannabinoid release, with distinct downstream consequences.

When calcium levels within the neuron increase or particular G-protein-coupled receptors are triggered, endocannabinoids are secreted non-synaptically and non-vesicularly from several locations throughout the cell. Endocannabinoids move to CB1R on neighboring neurons upon cellular release and cause a reversible, temporary reduction of synaptic communication, known as DSI. DSI causes a transitory decrease in GABA secretion from inhibitory interneuron presynaptic terminals in deactivated hippocampus pyramidal cells. Pharmacological experiments have provided the most direct information on the effects of endocannabinoids. Endocannabinoids' hydrophobicity prevents them from penetrating the tissue of the brain, and they are promptly degraded by numerous endogenous lipases. Because of these inherent characteristics of endocannabinoids, investigating their physiological impacts is difficult. These issues have been solved by the improvement of a greatly water-dissolvable caged AEA which is lipase inert (Ameri, 1999). The caged AEA acts as a dormant endocannabinoid pool if perfused into hippocampal slice preparations, and focused photolysis quickly releases extremely hydrophobic AEA *in situ* to trigger CB1R. By direct applying neuromodulators or NTs to synapses, circuits, or cells, photolysis is an additional practical way to excite synapses by chemically, circuits or cells.

Pharmacological techniques frequently provide limited control over-activation in the words of location, time, and uniqueness. The pharmaceutical technique has become more complex thanks to photo-uncaging of caged NTs. Photo-uncaging produces improved temporal and spatial resolution by utilizing targeted, patterned light. The flash photolysis approach is one use of photostimulation that may be utilized to measure the endocannabinoid system's signaling kinetics. The endocannabinoid system may be utilized to study the temporal dynamics and kinetics of lipid signaling. Endocannabinoid signaling kinetics may be determined using complete-cell voltage patch-clamp readings, intracellular Ca calculations, and light release of caged glutamate and a new caged cannabinoid, AEA. In this effort, flash photolysis of confined composites (photolysis employing as such molecular optical queries or photo probes) is a piece of useful equipment. Caged compounds (for example, caged glutamate or caged cannabinoid) are inert and physiologically inactive till a spark of laser light bursts apart the cage of the molecule, liberate the caged molecule and form a biologically dynamic effectors molecule in situ. The caged compound is a signing molecule that has been changed chemically. The molecule's alteration disables its bioactivity till light reuptake causes a photochemical variation in the signing molecule, restoring its bioactivity (Di Marzo et al., 1998).

The lipid signaling passageway has numerous temporal elements which may be measured to identify the time between the commencement of DSI and the onset of the DSI-inducing stimuli. Such factors contribute to the delay of DSI onset (starting of DSI-inducing stimulus to preliminary IPSCs repression). The rising of Ca stimulate endocannabinoid synthesis is one of them (t-Ca). Endocannabinoid production and release are triggered by a rise in intracellular Ca, and such molecules then go to CB1R on presynaptic interneurons, t-EC. The stimulation of CB1R and its descending consequences, t-CB1R (t-DSI = t-Ca + tEC + t-CB1R), are the following steps. Experiments utilizing the technological advances enabled researchers to determine the time it takes for eCB to be synthesized and released from the postsynaptic neuron, which had been approximated to be about 150 milliseconds at room temperature, which is a minimum array of magnitude that is faster than previously thought. The stimulation of presynaptic CB1Rs and the descending effects account for a large percentage of the DSI onset time, t-DSI. The findings show that endocannabinoid signaling is quick enough to regulate synaptic transmission at the moment, rather than just performing long-term neuromodulatory activities.

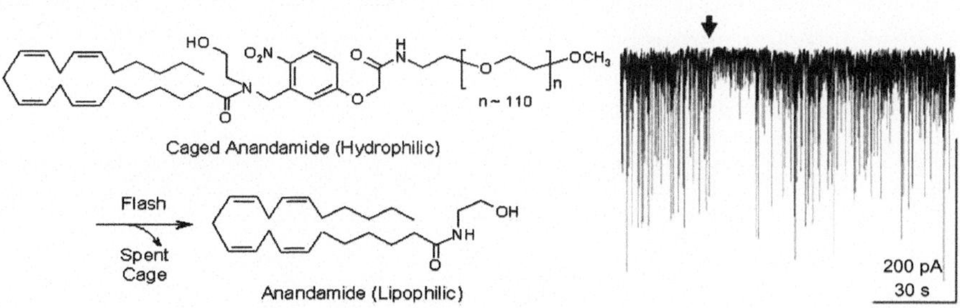

Figure 6.4. *Left-hand side panel: Photolysis of caged anandamide produces bioactive anandamide. Right-hand side panel: after a 180 milliseconds delay, photo-released anandamide decreases sIPSCs.*

Source: https://www.researchgate.net/publication/263662509_Neurochemical_Communication_The_Case_of_Endocannabinoids.

After a voltage step, the DSI onset delay, t-DSI, is 350 to 400 milliseconds. It takes around 180 milliseconds for t-CB1R, which is the straightforward stimulation of CB1R via photo secreting AEA, resulting in the inhibition of sIPSCs (Figure 6.4). A 50 milliseconds voltage step elicits a transitory rise in intracellular Ca adequate to attain minimum DSI, t-Ca, which requires about 60 milliseconds. The time it takes for endocannabinoid secretion and manufacturing, t-EC, is around 150 milliseconds (Devane et al., 1992).

An increase in intracellular concentration of calcium triggers voltage-activated DSI. Thus, even in the existence of a higher intracellular chelator [Ca], the secretion of endocannabinoids can be stimulated, but they can still be responsive to prevailing intracellular Ca (Mechoulam et al., 1995). The dynamical elements of the mGluR-induced endocannabinoid retaliation on sIPSC occurrence in pyramidal cells are examined (Figure 6.5) to check if these two routes function at a similar time scale. The IPSC repression elicited via uncased glutamate has the same onset latency, duration, and amplitude as that elicited by uncaged AEA (Figures 6.4 and 6.5). For 221 milliseconds, there was no drop in sIPSCs (calculated through extrapolation of the exponential fit to the regulated sIPSC level). The time to commencement of IPSC suppression mediated via the mGluR-driven endocannabinoid procedure (time to mGluR-dependent suppression of inhibition, tmGluRSI) is calculated as follows: t-mGluR-SI = 221 milliseconds = t-eCBs(mGluR) + t-CB1R, wherein teCB (mGluR) is the activation time of the mGluR-dependent endocannabinoid production and secretion, and t-CB1R is equivalent to 180 milliseconds. This leaves t-eCBs(mGluR) to be less than 50 milliseconds, which is quicker than endocannabinoid production and secretion activated through a voltage increase (Sugiura et al., 1995; Nicoll and Alger, 2004).

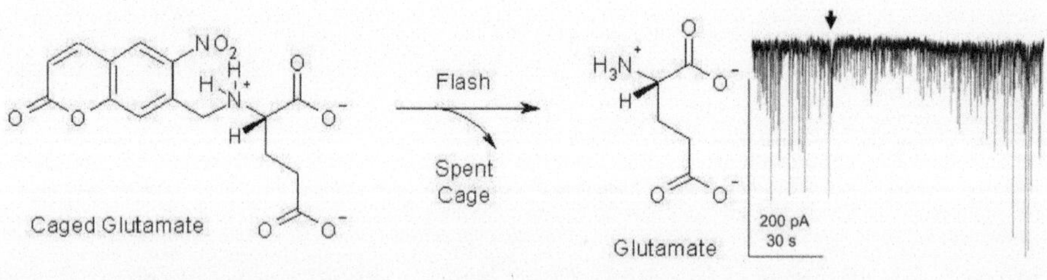

Figure 6.5. *In cultured hippocampal layers, there are dynamics of mGluR-dependent endocannabinoid suppression of sIPSC.*

Note: Photo release of the glutamate is on the left panel. Right-hand side panel: An information from a pyramidal cell shows a transitory drop in CA1 pyramidal cells' spontaneous (s) IPSC frequency while caged glutamate was a flash photo released (photolysis induced repression of inhibition, PSI). The arrow denotes a laser flash (Maejima et al., 2001).

Source: https://www.researchgate.net/publication/263662509_Neurochemical_Communication_The_Case_of_Endocannabinoids.

6.6. DSI IN GLOMERULAR PATHWAYS OF THE OLFACTORY BULB

The olfactory bulb is the initial convey station in the central nervous system (CNS) for processing sensory data from the nostril epithelium's olfactory receptor cells. The olfactory bulb, notably the input portion, the glomerular layer, has large amounts of CB1R (Ohno-Shosaku et al., 2001; Varma et al., 2001). Enzymes which produce endocannabinoids are immunoreactive in neurons in the glomerular layer. The physiological significance of cannabinoid and endocannabinoids receptors in neuronal signaling in the olfactory system is only beginning to be understood. The endocannabinoid system performs a practical part in modulating neuronal action and signaling in the olfactory bulb glomeruli, according to current electrophysiological data (Wilson and Nicoll, 2001).

The glomerular neuronal population is subdivided into three subpopulations: short-axon (SA), external tufted (eTC), and periglomerular (PG). Functionally and neurochemically, the periglomerular cells are highly diverse. External tufted cells are glutamatergic, periglomerular cells are GABAergic, short-axon cells contain both DA and GABA, and periglomerular cells are GABAergic. Periglomerular cells get olfactory nerve or dendrodendritic glutamatergic addition via exterior mitral or tufted cells, such as EPSCs. By GABAergic communication, periglomerular cells presynaptically suppress olfactory receptor neurons. Exterior tufted cells get sIPSCs and also spontaneous glutamatergic EPSCs from periglomerular cells across inhibitory GABAergic synapses. Exterior tufted cells in the glomerular film may be a resource of endocannabinoids (Wilson et al., 2001; Kreitzer and Regehr, 2001).

The impacts of agonist WIN and CB1R antagonist AM251 in the existence of GABAA receptor inhibitors (synaptic inhibitors: gabazine, APV, CNQX,) and ionotropic glutamate reveal that CB1R significantly control membrane characteristics of periglomerular cells (Maejima et al., 2001). It suggests that cannabis' effects on periglomerular cells have been mediated via CB1R, which is represented by periglomerular cells. AM251 stimulates periglomerular cells directly and increases their GABA secretion. External tufted cells are synaptically related to periglomerular cells. Synaptic transmission and GABA release to outer tufted cells might be affected by every CB1R-mediated control of periglomerular cell function. Exterior tufted cells possess CB1R, which can play a role in modifying exterior tufted cell function (Figure 6.6) (Yoshida et al., 2002).

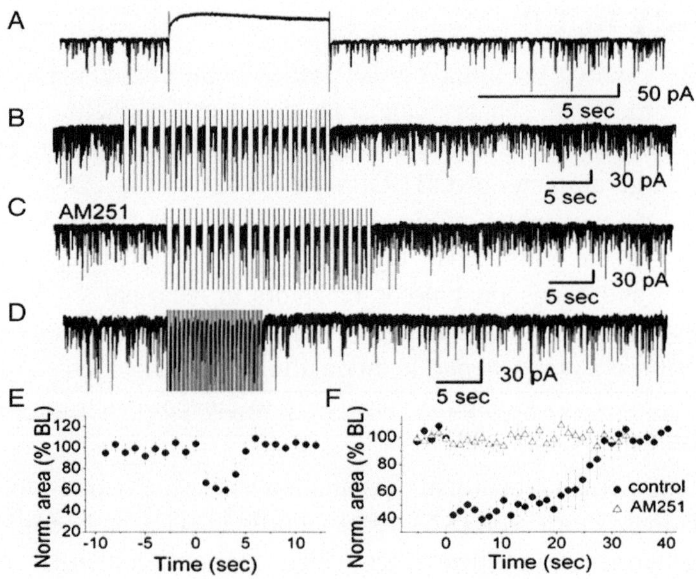

Figure 6.6. *In olfactory glomeruli, there is depolarization-induced suppression of inhibition (DSI).*

Note: (A) In a typical exterior tufted cell, a step of depolarizing voltage elicited DSI. For recording sIPSCs, a higher chlorine-based pipette solution was employed. Stepping from a –60 millivolt holding potential to zero millivolt for five seconds resulted in depolarization. (B) A train of 20 voltage steps to zero millivolt (0.75 Hertz; period of step: 667 milliseconds) transiently suppressed sIPSCs in an exterior tufted cell in the existence of CNQX and 5-AP. The holding potential was set at –60 mV. (C). There was no sIPSC suppression in the presence of AM251. (D) In an exterior tufted cell, a train of 20 voltage steps to –30 mV (2 Hertz; period of step: 250 milliseconds) transiently suppressed sIPSCs (in CNQX and 5-AP). (E) The time and magnitude history of DSI induced by a five-second depolarizing pulse (n=7) are depicted by normalized sIPSCs area. The averaged values b between zero and five seconds after the endpoint of voltage step had been substantially distinct from the baseline (ANOVA and

Bonferroni analysis of post-hoc, p is less than 0.05). (F) Standardized sIPSC area depicting the time and magnitude course of DSI induced by a depolarizations train to zero millivolts (n=12) in control and AM251 (n=10) in the existence of AM251. The mean values between 0 and 25 sec. after the conclusion of the voltage step train were substantially distinct from the baseline under control circumstances (ANOVA and Bonferroni analysis of post-hoc, p is less than 0.05) (Trettel and Levine, 2003; Trettel et al., 2004).

Source: https://onlinelibrary.wiley.com/doi/10.1111/ejn.15186.

REFERENCES

1. Alger, B. E., & Kim, J., (2011). Supply and demand for endocannabinoids. *Trends Neurosci., 34*, 304–315.
2. Alger, B. E., (2001). Endocannabinoids at the synapse a decade after the dies mirabilis (29 March 2001): What we still do not know. *J. Physiol., 590*(10), 2203–2212.
3. Alger, B. E., (2002). Retrograde signaling in the regulation of synaptic transmission: Focus on endocannabinoids. *Prog. Neurobiol., 68*, 247–286.
4. Ameri, A., (1999). The effects of cannabinoids on the brain. *Prog. Neurobiol., 58,* 315–348.
5. Calignano, A., Katona, I., Desarnaud, F., Giuffrida, A., La Rana, G., Mackie, K., Freund, T. F., & Piomelli, D., (2000). Bidirectional control of airway responsiveness by endogenous cannabinoids. *Nature, 408*, 96–101.
6. Cota, D., Marsicano, G., Lutz, B., Vicennati, V., Stalla, G. K., Pasquali, R., & Pagotto, (2003). Endogenous cannabinoid system as a modulator of food intake. *Internat. J. Obesity, 27*, 289–301.
7. Devane, W. A., Hanus, L., Breuer, A., Pertwee, R. G., Stevenson, L. A., Griffin, G., et al., (1992). Isolation and structure of a brain constituent that binds to the cannabinoid receptor. *Science, 258,* 1946–1949.
8. Di Marzo, V., Melck, D., Bisogno T., & De Petrocellis, L., (1998). Endocannabinoids: Endogenous cannabinoid receptor ligands with neuromodulatory action. *Trends Neurosci., 21*, 521–528.
9. Freund, T. F., Katona, I., & Piomelli, D., (2003). Role of endogenous cannabinoids in synaptic signaling. *Physiol. Rev., 83*, 1017–1066.
10. Herkenham, M., Lynn, A. B., Johnson, M. R., Melvin, L. S., De Costa, B. R., & Rice, K. C., (1991). Characterization and localization of cannabinoid receptors in rat brain: A quantitative *in vitro* autoradiographic study. *J. Neurosci., 11*, 563–583.
11. Herkenham, M., Lynn, A. B., Little, M. D., Johnson, M. R., Melvin, L. S., De Costa, B. R., & Rice, K. C., (1990). Cannabinoid receptor localization in brain. *Proc. Natl. Acad. Sci. USA, 87*, 1932–1936.
12. Howlett, A. C., Breivogel, C. S., Childers, S. R., Deadwyler, S. A., Hampson, R. E., & Porrino, L. J., (2004). Cannabinoid physiology and pharmacology: 30 years of progress. *Neuropharmacol, 47*, 345–358.
13. Iversen, L., & Chapman, V., (2002). Cannabinoids: A real prospect for pain relief. *Curr Opin Pharmacol., 2*, 50–55.
14. Kano, M., Ohno-Shosaku, T., Hashimotodani, Y., Uchigashima, M., & Watanabe, M., (2009). Endocannabinoid-mediated control of synaptic

transmission. *Physiol. Rev., 89*, 309–380.

15. Katona, I., Rancz, E. A., Acsady, L., Ledent, C., Mackie, K., Hajos, N., & Freund, T. F., (2001). Distribution of CB1 cannabinoid receptors in the amygdala and their role in the control of GABAergic transmission. *J. Neurosci., 21*, 9506–9518.
16. Katona, I., Sperlagh, B., Sik, A., Kafalvi, A., Vizi, E. S., Mackie, K., & Freund, T. F., (1999). Presynaptically located CB1 cannabinoid receptors regulate GABA release from axon terminals of specific hippocampal interneurons. *J. Neurosci., 19*, 4544–4558.
17. Kreitzer, A. C., & Regehr, W. G., (2001). Cerebellar depolarization-induced suppression of inhibition is mediated by endogenous cannabinoids. *J. Neurosci., 21*, RC174.
18. Maejima, T., Hashimoto, K., Yoshida, T., Aiba, A., & Kano, M., (2001). Presynaptic inhibition caused by retrograde signal from metabotropic glutamate to cannabinoid receptors. *Neuron., 31*, 463–475.
19. Maejima, T., Ohno-Shosaku, T., & Kano, M., (2001). Endogenous cannabinoid as a retrograde messenger from depolarized postsynaptic neurons to presynaptic terminals. *Neurosci. Res., 40*, 205–210.
20. Marsicano, G., Goodenough, S., Monory, K., Hermann, H., Eder, M., Cannich, A., Azad, S. C., et al., (2003). CB1 cannabinoid receptors and on-demand defense against excitotoxicity. *Science, 302*, 84–88.
21. Marsicano, G., Wotjak, C. T., Azad, S. C., Bisogno, T., Rammes, G., Cascio, M. G., Hermann, H., et al., (2002). The endogenous cannabinoid system controls extinction of aversive memories. *Nature, 418*, 530–534.
22. Matsuda, L. A., Bonner, T. I., & Lolait, S. J., (1993). Localization of cannabinoid receptor mRNA in rat brain. *J. Comp. Neurol., 327*, 535–550.
23. Matsuda, L. A., Lolait, S. J., Brownstein, M. J., Young, A. C., & Bonner, T. I., (1990). Structure of a cannabinoid receptor and functional expression of the cloned cDNA. *Nature, 346*, 561–564.
24. Mechoulam, R., Ben-Shabat, S., Hanus, L., Ligumsky, M., Kaminski, N. E., Schatz, A. R., Gopher, A., et al., (1995). Identification of an endogenous 2-monoglyceride, present in canine gut, that binds to cannabinoid receptors. *Biochem. Pharmacol., 50*, 83–90.
25. Munro, S., Thomas, K. L., & Abu-Shaar, M., (1993). Molecular characterization of a peripheral receptor for cannabinoids. *Nature, 365*, 61–65.
26. Nicoll, R., & Alger, B. E., (2004). The brain's own marijuana. *Sci. Amer., 291*, 68–75.

27. Núñez, E., Benito, C., Pazos, M. R., Barbachano, A., Fajardo, O., González, S., et al., (2004). Cannabinoid CB2 receptors are expressed by perivascular microglial cells in the human brain: An immunohistochemical study. *Synapse, 53*, 208–213.
28. Ohno-Shosaku, T., Maejima, T., & Kano, M., (2001). Endogenous cannabinoids mediate retrograde signals from depolarized postsynaptic neurons to presynaptic terminals. *Neuron, 29*, 729–738.
29. Panikashvili, D., Simeonidou, C., Ben-Shabat, S., Hanus, L., Breuer, A., Mechoulam, R., & Shohami, E., (2001). An endogenous cannabinoid (2-AG) is neuroprotective after brain injury. *Nature, 413*, 527–531.
30. Soria-Gómez, E., Bellocchio, L., Reguero, L., Lepousez, G., Martin, C., Bendahmane, M., Ruehle, S., et al., (2014). The endocannabinoid system controls food intake via olfactory processes. *Nat. Neurosci., 17*, 407–415.
31. Sugiura, T., Kondo, S., Sukagawa, A., Nakane, S., Shinoda, A., Itoh, K., Yamashita, A., & Waku, K., (1995). 2-Arachidonoylgylcerol – a possible endogenous cannabinoid receptor-ligand in brain. *Biochem. Biophys. Res. Commun., 215*, 89–97.
32. Trettel, J., & Levine, E. S., (2003). Endocannabinoids mediate rapid retrograde signaling at interneuron pyramidal neuron synapses of the neocortex. *J. Neurophys., 89*, 2334–2338.
33. Trettel, J., Fortin, D. A., & Levine, E. S., (2004). Endocannabinoid signaling selectively targets perisomatic inhibitory inputs to pyramidal neurones in juvenile mouse neocortex. *J. Physiol. (London), 556*, 95–107.
34. Tsou, K., Mackie, K., Sanudo-Pena, M. C., & Walker, J. M., (1999). Cannabinoid CB1 receptors are localized primarily on cholecystokinin-containing GABAergic interneurons in the rat hippocampal formation. *Neurosci., 93*, 969–975.
35. Valjent, E., Mitchell, J. M., Besson, M. J., Caboche, J., & Maldonado, R., (2002). Behavioral and biochemical evidence for interactions between Delta9-tetrahydrocannabinol and nicotine. *Br. J. Pharmacol., 135,* 564–578.
36. Van, S. M. D., Duncan, M., Kingsley, P. J., Mouihate, A., Urbani, P., Mackie, K., et al., (2005). Identification and functional characterization of brainstem cannabinoid CB2 receptors. *Science, 310,* 329–332.
37. Varma, N., Carlson, G. C., Ledent, C., & Alger, B. E., (2001). Metabotropic glutamate receptors drive the endocannabinoid system in the hippocampus. *J. Neurosci., 21*(RC188), 1–5.
38. Wilson, R. I., & Nicoll, R. A., (2001). Endogenous cannabinoids mediate retrograde signaling at hippocampal synapses. *Nature, 410,* 588–592.

39. Wilson, R. I., Kunos, G., & Nicoll, R. A., (2001). Presynaptic specificity of endocannabinoid signaling in the hippocampus. *Neuron, 31*, 1–20.
40. Yoshida, T., Hashimoto, K., Zimmer, A., Maejima, T., Araishi, K., & Kano, M., (2002). The cannabinoid CB1 receptor mediates retrograde signals for depolarization-induced suppression of inhibition in cerebellar Purkinje cells. *J. Neurosci., 22*, 1690–1697.

CHAPTER

7 Behavioral Neurochemistry

Contents

7.1. INTRODUCTION

One of the most important contributions emerging from Skinner's pioneering work almost 50 years before has been the discovery and implementation of experimental procedures to examine ongoing behavior (Skinner, 1938). Skinner's theory and research sparked an intense study of operant behavior, allowing for detailed resolution of behavior and discovery of its regulating factors. Complicated aspects of behavior and the various variables that affect their temporal distribution and appearance have become readily available thanks to the employment of very simple approaches. Numerous basic principles of behavior have developed, with far-reaching implications and applications far outside the realm of behavioral science. Skinner (1938), who was exclusively engaged in the scientific study of human behavior at the time, noticed that behavior may have certain 'neurological' correlations and even recommended how the sciences that merged behavior with a 'Physicochemical' approach could best progress.

It is reasonable to suppose that a science of the nervous system would eventually begin with direct observation of brain procedures and develop ideas and rules in response. The neurological perspective should be concerned with this research if it is to provide a credible "explanation" for behavior (Mansbach et al., 1988).

Skinner seems to be particularly concerned at the time with the lack of acceptable procedures for directly and consistently assessing variations in central nervous system (CNS) activity. This deficiency, although, was not thwarted, tries to explain behavior using unmeasured neural procedures. Whereas many participants of the experimental study of human behavior have shown a preference for domains besides behavior and its surroundings, this preference appears to come in part from a period when CNS variables were not easily available or objectively recognizable. Accessibility to dynamic CNS activities capable of performing simultaneous studies of the neurochemical and behavioral impacts of, say, reinforcement contingencies has been exceedingly unpleasant and complex. However, when strategies for gaining access to CNS activities were established, they were frequently limited to static windows that opened for a small period. As is generally known, behavior evolves through time and is a product of a variety of past and current circumstances. An appreciation and comprehension of the reciprocal impacts between neurochemistry and behavior should be capable to capture what must be an enthrallingly dynamic interaction (Marsden, 1985).

Many people realized that success in the experimental study of human behavior did not have to wait for breakthroughs in other studies since principles formulated at the level of behavior will stay consistent regardless of what happened in other disciplines. Much notable advancement in similar fields has been made in recent years, making the research of CNS correlates of behavior considerably more practical for those who are interested (Figure 7.1) (Marsden et al., 1988; Miyauchi et al., 1988).

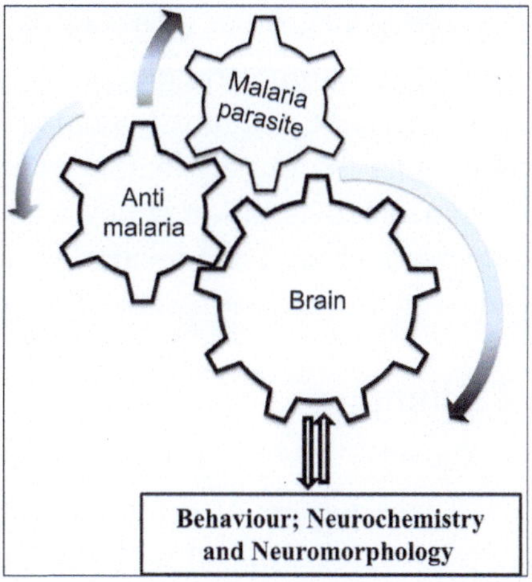

Figure 7.1. *Elements of behavior neuromorphology and neurochemistry.*

Source: https://benthamscience.com/journals/central-nervous-system-agents-in-medicinal-chemistry/volume/18/issue/3/.

More modern approaches provide an overview of neurochemical interactions and behavioral that comes close to the resolution required and is long enough to promote progress in this sector. Methods nowadays are available to assess neurochemical processes consistently and directly, and these events may now be correlated with behavior. Neuropharmacological and neurochemical assessments are sometimes as measurable and objective as behavioral analyzes. They may be virtually immediate, giving the experimenter the sort of reward that comes from seeing modifications in ongoing behavior as a result of environmental manipulations. The goal of investigating these processes is to establish the impact of behavioral factors and behavior on neurochemistry, rather than to 'explain' behavior. Such attempts do not lessen the relevance of factors that regulate behavior. Methods that provide access to various variables to provide a more complete examination of behavior may only expand and explain fundamental behavioral procedures (Moore, 1978; Moghaddam and Bunney, 1989).

7.2. FUNDAMENTAL PROCEDURES

Before attempting any of the methods discussed, it is necessary to have a thorough understanding and proficiency with several fundamental techniques that are usually accessible in most physiological psychology laboratories. Almost every technique and methodology that requires the usage of a stereotaxic tool to precisely locate and put cannulae or probe instruments utilized in the collection of fluid, injection of the drug, or the insertion of a monitoring electrode into particular areas of the brain. The majority of the methods include fluid collection from particular areas of the brain in awake, behaving animals, and they may be technically demanding. Animals should be observed considerably more thoroughly than is common or conventional in most behavioral experiments, in addition to basic surgical skills, usage of a stereotaxic device, maintenance, and administration of anesthesia (Morley et al., 1987a, b).

7.2.1. Stereotaxic Techniques

Pellegrino and Cushman (1971) discussed the usage of stereotaxic devices and stereotaxic atlases. These writers also include a list of other sources for stereotaxic atlases that may be utilized with a variety of species. Once the study is over, appropriate histological ways of evaluating and finding the areas for cannula placement, the level of damage, and so on should be accessible. Concerning the Cushman and Pellegrino article, numerous specialized texts cover such subjects in greater depth and must be reviewed before commencing on the investigation of nature (Bures et al., 1983; Joseph and Waddington, 1986).

7.2.2. Anesthesia

Most anesthetic maintenance and induction protocols have been refined to offer adequate detail for procedures to be carried out compassionately, successfully, and safely. Waddington (1986) offered a detailed account of the process for inducing anesthesia in rats, as well as the fundamental procedure for doing stereotaxic surgery on rats. The majority of rat studies employ barbiturate anesthesia (for example, 50 to 60 mg/kg sodium pentobarbital delivered intraperitoneally), while primates utilized an inhalation anesthetic like halothane (2% oxygen). Rats are frequently administered atropine sulfate (0.05 mg/kg, s.c.) like a pre-anesthetic to reduce salivary and bronchial discharges and avoid respiratory discomfort during surgery. In pigeons, an i.m. solution of 10 mg/kg pentobarbital sodium and 30 to 40 mg/

kg ketamine hydrochloric acid produces better levels of anesthesia which are harmless; subsequent injections of 10 mg/kg ketamine would sustain these levels with no toxicity. To reduce trauma to the animal and the gadgets that have been implanted, recovery from anesthetic must be slow and gradual. Anesthesia recovery must be allowed in a clean and warm environment. Likewise, proper caging is required to prevent the animal from destroying probes, injury, or cannulae. This can entail keeping the animals in cages without wire mesh or in bigger cages to prevent the probes or cannula from becoming caught and being dragged out (Figure 7.2) (Peterson and Sparber, 1974; Pettit and Justice, 1989).

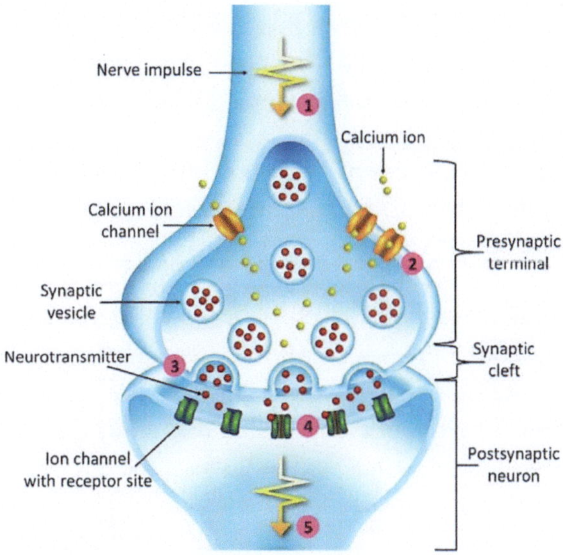

Figure 7.2. *Principle of the functioning of anesthesia.*

Source: https://neurosciencenews.com/general-anesthesia-mechanism-16259/.

7.2.3. Surgical Preparation and Procedure

Appropriate surgical processes that pay special attention to the animal's care before surgery, the choice of proper surgical tools, sterilization processes, and the animal's preparation for surgery all make a significant contribution to the study's achievement. When working with nonhuman primates, sterile surgical methods are required; however, this is not required when working with most regularly utilized species, like rats or pigeons. Surgical tools, on the other hand, must be kept in good working order and kept clean, as is clear (Pycock and Taberner, 1981; Porrino et al., 1984). A mixture of pure alcohol (500 mL) and 19% zephiran (7 mL) combined with distilled water appears to be a particularly effective 'cold sterilization' procedure (500 ml). Instruments are soaked for 30 minutes before operation and then cleaned in sterile distilled water before use. Numerous articles and texts on small-animal surgery are available, and they give excellent overviews of the stages involved in such techniques (Meyer and Meyer, 1971).

7.2.3.1. Preparation of the Skull and Insertion of the Cannula

The animal may be positioned in the stereotaxic device after surgical anesthesia has been attained and the skin region in which the incision will be made has been clipped or shaved and washed with betadine or a suitable equivalent. The exact placement of the incision is determined by the location of the cannula or probe device, but it should be big enough to allow adequate tissue reflection to provide unrestricted access to the desired spot. Retractors are used to keep the skin open. To secure the cranioplastic cement that supports the cannula, three to four stainless steel screws would be put into the skull in most cases. As a result, the periosteum should be scraped away, and the cement-covered surface of the skull should be dry and clean at the time of preparation. Most cement retention screws are self-tapping and may be readily inserted once the original hole has been drilled with a sufficient size drill bit. The stainless-steel screws must be threaded down far enough to create a firm anchor without compressing the brain; also, the screw head's base must not rest directly on the skull, allowing the cranioplastic cement to flow below and offer appropriate support (Sakurai-Yamashita et al., 1989; Salamone et al., 1989).

The hole for the cannula may be formed once the retaining screws are inserted and tightened. This type of hole is usually opened using a dental drill that is used with caution to prevent damaging the surrounding tissue or bone. Opening a big enough region to allow for visualization of the brain surface, exposure, and puncture of the dura, and proper insertion of the cannula with no compression or distortion of the brain are all desirable outcomes. Following placement of the cannula at the appropriate depth, cranioplastic cement is administered to hold the cannula in position and protect it throughout the following experiments. Initially, the cement should be mixed in such a way that it flows freely around and under the mounting screws; the following applications may be more viscous and may be applied after every preceding application has been allowed to dry completely. There are various advantages to transparent cement over opaque cement, including the fact that it is easier to monitor for bleeding beneath the cement and that it may be quickly removed, if necessary, without impairing the vision of the skull or surrounding tissue while drilling out old cement. Since the hardening of the cement creates heat and may cause tissue damage and pain, it is important to avoid direct contact with the cement with the skin drying (Figure 7.3) (Schallert and Wilcox, 1985).

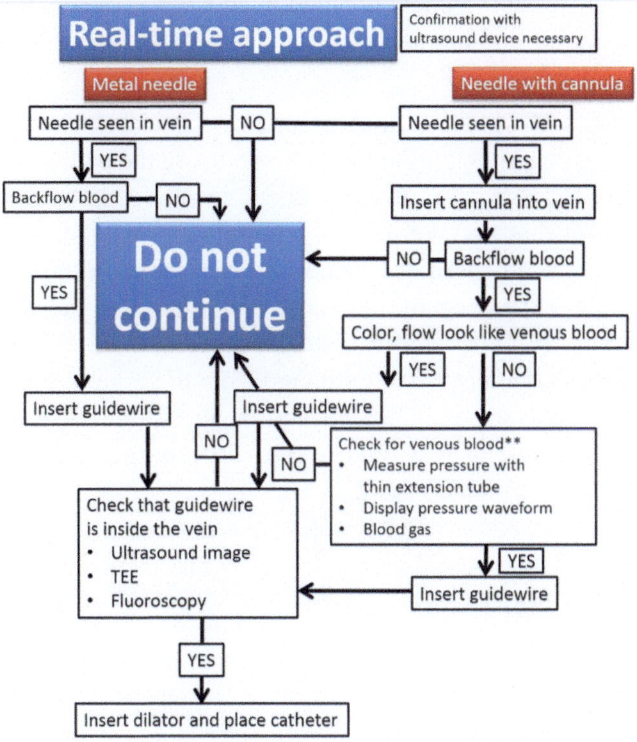

Figure 7.3. *Techniques of central venous catheterization using a static approach.*

Source: https://link.springer.com/article/10.1007/s00540-019-02702-9.

7.2.4. Analyzing Behavior: Technological Factors, Issues Detection and Minimization of Problems

Many of the experiments discussed briefly need particular adaptations to the equipment to avoid injuries to the guide cannula, tubing attached to withdrawal pumps, and other components. The majority of these changes are self-evident and do not need further explanation. Pigeons with a guiding cannula glued to the skull, for instance, require a bigger feeding aperture than most commercially constructed grain magazines may give. The hole should be extended to a minimum height of nine centimeters so that the cannula does not reach the top of the opening when the pigeon inserts its head into it. It is important to utilize a feeding cup positioned on the front panel of the chamber, immediately below the response buttons, when cerebrospinal fluid (CSF) or extracellular fluids are being extracted constantly throughout an experimental session. Pellets for pigeon checkers can be administered in the same way as rats and monkeys get their pellets (Schoenfeld and Uretsky, 1972; Schwarcz et al., 1978).

Many efforts to obtain a fluid specimen from acting animals have a fundamental challenge: retaining the tubing outside the path and unavailable to the animal. The majority of researchers employ a balancing arm with an attached pivot that may be modified to move in reaction to the animal. There must be certain sacrifices made between maintaining the necessary isolation of the experimental situation and allowing the specimens to be collected swiftly and without disturbing the animal. Several NTs and metabolites deteriorate fast when extracted from the brain through push-pull perfusion, microdialysis, and CSF extraction, and should be preserved from deterioration as rapidly as feasible (Figure 7.4) (Sharp et al., 1984, 1986).

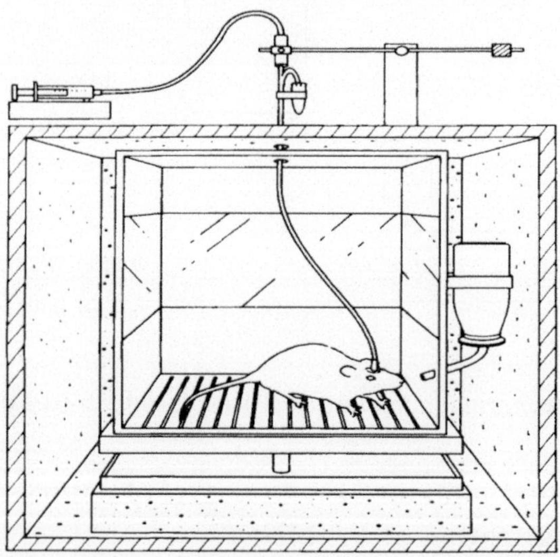

Figure 7.4. *A rat is shown which is linked to a balancing arm and swivel for unrestricted dialysate collection from a microdialysis probe. Dialysis specimens were collected in the microcentrifuge tube at the top of the chamber, outside of the chamber (Kuczenski and Segal, 1989).*

Source: https://pubmed.ncbi.nlm.nih.gov/2566664/.

As a result, it is critical to keep the length of tubing used to collect specimens as short as possible while maintaining the integrity of the experimental environment so that behavior is not disrupted. The gap between the top of the experimental chamber and the inner ceiling of the surrounding acoustical enclosure may be minimized to achieve this. The tube may then depart the experimental chamber and travel only a small distance to the pump or collecting device, which is conveniently situated right on top of the acoustical chamber (Sparber, 1975; Soubrie, 1986).

7.2.5. High-Performance Liquid Chromatography (HPLC) with Electrochemical Detection

Using the technology of high-performance liquid chromatography (HPLC) in conjunction with an electrochemical monitoring technique, it is possible to determine pictogram or smaller levels of a broad variety of behaviorally and biologically active chemicals present in fluid or tissue. Another significant benefit is that many substances may be analyzed at the same time from a single specimen, allowing for the most efficient utilization of specimen resources and animal resources (Thoenen and Tranzer, 1968).

Although some substances, like indoles and phenols, had been recognized for a certain time to be electrochemically active, it was not until the early 1970s that Adams and his colleagues realized that the oxidation of tyrosine and tryptophan metabolites could allow the quantification of such substances in the tissues of brain (Adams and Marsden, 1982). A chromatographic method, HPLC is used to separate different chemical species in a mixture. However, unlike the other forms of separation (such as gas and paper chromatography), HPLC makes usage of the liquid mobile process which is driven by a pump through an analytical column packed firmly with tiny particles; this is in contrast to other forms of separation (such as gas and paper chromatography). The size and type of packaging material, along with column size, and also multiple other variables, including the specific composition of the mobile phase (polarity, pH, etc.), are essential in determining the time required for biological substances embedded in the mobile phase to elute or pass-through from the column (retention time) and traverse the electrochemical detector. The electrochemical cell is often a thin-layer sandwich-type configuration in which the volume at the detecting region is quite tiny, as is the case with most semiconductor devices (Tilson and Sparber, 1970; Tye et al., 1977).

Here, on the surface of an electrode, biologically electroactive substances that are introduced into the mobile phase are oxidized, and the current generated at this point is enhanced and transformed into a voltage that is then recorded on a strip chart or other kind of integrating or recording device. Figure 7.5 is an illustration of a chromatogram obtained from a study of the CSF of a squirrel monkey following either saline administration or treatment of the serotonin precursor L-5-hydroxytryptophan (L-5-HTP). The peaks clearly show the elution of the metabolites of DA and serotonin in the correct sequence. Using recognized standards that are tested in a similar system, it is possible to determine whether or not these chemicals are identified. Metabolite and NT evaluation using HPLC with electrochemical detection is a difficult and technically demanding procedure (Ungerstedt and Hallström, 1987; Uhl, 1988).

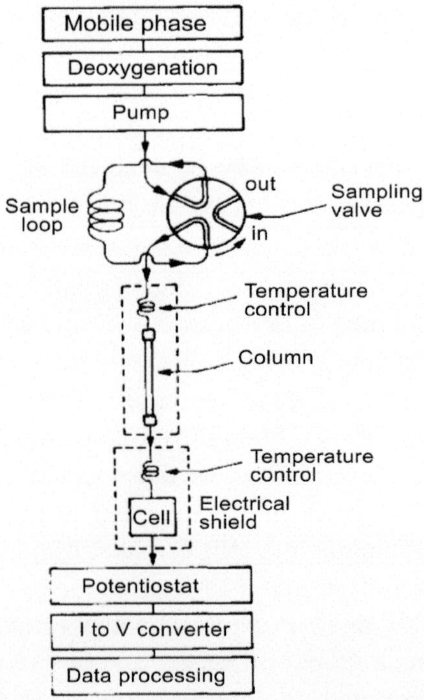

Figure 7.5. *Chromatogram of CSF collected from a squirrel monkey under control (injection of saline) circumstances and after the L-5-hydroxytryptophan (5-HTP) administration. The graph depicts the elution pattern of the metabolites.*

Source: https://www.semanticscholar.org/paper/Chapter-2-%E2%80%93-Determination-of-biogenic-amines%2C-their-Duda-Kissinger/136f5893b177acd7ef8c979d632837da38c4222e.

7.3. PUSH-PULL PERFUSION AND CEREBROSPINAL FLUID (CSF) WITHDRAWAL

Because of their widespread usage in physiology since their debut in the late 1950s, push-pull perfusion methods have been studied extensively. Early investigations, which were described by Myers and Knott (1986), comprised of inserting a side-by-side arrangement of the tubing into the forearm of a man subject to retrieve bradykinin. Perfusion medium (such as saline or solution of Ringer) was injected into the area through one piece of tubing, whereas the perfusate (which contained the perfusion solution as well as whatever physiological substances had been accessible at the location being perfused) was withdrawn through the other piece of tubing (Hilton and Fox, 1958). It was Gaddum (1961) who developed a modified version of this approach to create a concentric cannula that could deliver more targeted perfusion and made the 1st observation of the release of acetylcholine (ACh) from the caudate nucleus of the brain (Gaddum, 1962). Almost immediately after its development, this method was put to widespread usage.

Figure 7.6 illustrates a typical concentric cannula design. The exterior guide tube is often permanently installed in the desired location and sealed with a 'dummy cannula' made up of a cap and styles (Plastic items Inc., Roanoke, VA 24022, offers the cannula, cannula guide, and caps and it would custom construct the assemblies). The longer component is put into the guide tube such that it stretches further than the endpoint of the guide during testing. When the push-pull perfusion starts, artificial CSF or suitable perfusion fluid is given into the push tubing, where it departs and mixes with endogenous neuronal fluid before being aspirated into the outside ('pull') tube, which goes to a collecting device. Two pumps or syringe drivers are required, both of which must function in tandem and also be calibrated to infuse and remove at the same pace. Myers (1972, 1974, 1977) wrote extensively on the techniques for doing push-pull tests and continues to innovate with this method. For instance, a new micro push-pull catheter system (Myers and Gurley-Orkin, 1985) and a contemporary double-lumen polyethene cannula (Myers et al., 1985) have been reported, both of which have benefits over the more traditional push-pull method. Although, it seems that little further work is done with such approaches as of yet, which might be related to the new introduction of microdialysis treatments. Many factors have a role in influencing the rate of recovery, the amount of perfused area, and the amount of tissue injury. Most notably, the pace at which the region is perfused and the length of the push tube extension further than the outer pull tubing are assessed extensively in model experimentations and in vivo studies (Szerb, 1967; Yaksh and Yamamura, 1974).

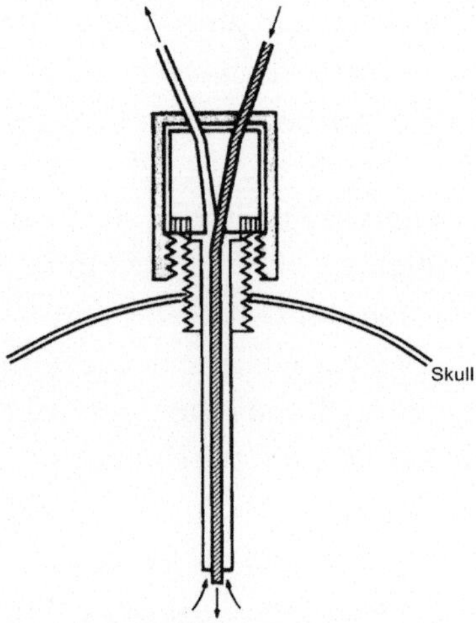

Figure 7.6. *The diagram of the push-pull cannula assembly.*

Source: https://www.sciencedirect.com/science/article/abs/pii/B9780444812513500102.

7.3.1. Measurement of Endogenous Neurotransmitters (NTs) and Metabolites

Amongst the most common applications of the push-pull perfusion approach is the collection of NTs and their metabolites from the ventricular space or regional regions of the brain following systemic drug administration. Furthermore, this treatment may be performed on animals that are awake and free to move around, allowing for a more effective evaluation of neuronal activity in geographically specific parts of the brain than will be feasible, for instance, under general anesthesia. Even though this treatment allows for repeated sampling from a similar animal, it is only permitted a limited number of times due to the creation of scar tissue, encapsulation of the cannula guide tip, and cellular disintegration. When integrated with the strong methods of HPLC and electrochemical identification, the push-pull perfusion methodology became more popular since studies might be performed with microliter volumes of perfusate (Nielsen and Moore, 1982). To collect solely CSF from the ventricular space or the cisterna magna, some researchers have improved the push-pull approach, once again focusing on the use of repeated sample procedures for long-term assessments of neurochemical activity (e.g., Sarna et al., 1983; De La Riva and Yeo, 1985; Barrett and Nader, 1990). These procedures involve implanting the guide cannula in or above the intended spot, which is subsequently pierced or tapped to collect CSF. In most circumstances, CSF is removed slowly by -ve pressure, and in certain cases, the system is left open to allow for the least amount of variation in ventricular pressure. This method of monitoring neurochemical activity by measuring metabolites in CSF causes the least amount of stress to any single brain location and may be sustained for significantly longer durations (Figure 7.7).

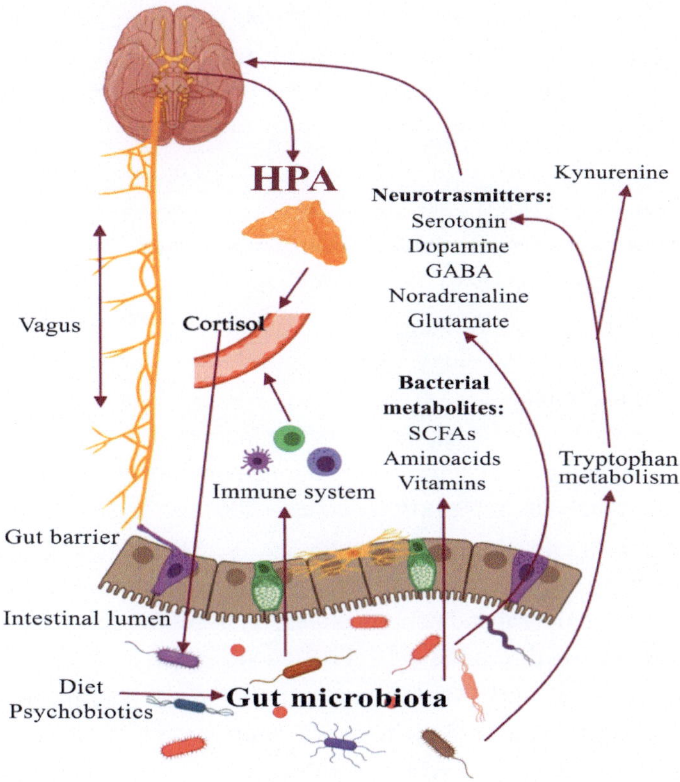

Figure 7.7. *The gut microbiome's function in the gut-brain axis is depicted in the diagram.*

Source: https://www.mdpi.com/1422-0067/21/23/9234/htm.

REFERENCES

1. Adams, R. N., & Marsden, C. A., (1982). Electrochemical detection methods for monoamine measurements in vitro and in vivo. In: Iversen, L.L., Iversen, S. D., & Snyder, S. H., (eds.), Handbook of Psychopharmacology (Vol. 15, pp. 1–74). Plenum Press, New York.

2. Barrett, J. E., & Nader, M. A., (1990). Neurochemical correlates of behavioral processes. Drug Dev. Res., 20, 313–335.

3. Bures, J., Buresovä, O., & Huston, J., (1983). Techniques and Basic Experiments for the Study of Brain and Behavior (2nd edn., Vol. 1, pp. 1–22). Elsevier, Amsterdam.

4. De La Riva, C. F., & Yeo, J. A. G., (1985). Repeated determination of cerebrospinal fluid amine metabolites by automated direct sampling from an implanted cannula in freely moving rats. J. Neurosci. Methods, 14, 233–240.

5. Gaddum, J. H., (1961). Push-pull cannulae. J. Physiol., 155, 1, 2P.

6. Gaddum, J. H., (1962). Substances released in nervous activity. In: Paton, W. D. M., & Lindgren, P., (eds.), Pharmacological Analysis of Central Nervous Action (pp. 1–6). Pergamon Press, London.

7. Joseph, M. H., & Waddington, J. L., (1986). Working Methods in Neuropsychopharmacology (Vol. 1, pp. 1–22). Manchester University Press, Manchester.

8. Kuczenski, R., & Segal, D., (1989). Concomitant characterization of behavioral and striatal neurotransmitter response to amphetamine using *in vivo* microdialysis. J. Neurosci., 9, 2051–2065.

9. Mansbach, R. S., Harrod, C., Hoffmann, S. M., Nader, M. A., Lei, Z., Witkin, J. M., & Barrett, J. E., (1988). Behavioral studies with anxiolytic drugs. V. Behavioral and neurochemical analyses in pigeons of drugs that increase punished responding. J. Pharmacol. Exp. Ther., 246, 114–120.

10. Marsden, C. A., (1985). In vivo monitoring of pharmacological and physiological changes in endogenous serotonin release and metabolism. In: Green, A. R., (ed.), Neuropharmacology of Serotonin (pp. 218–252). Oxford University Press, London.

11. Marsden, C. A., Joseph, M. H., Kruk, Z. L., Maidment, N. T., O'Neill, R. D., Schenk, J. O., & Stamford, J. A., (1988). In vivo voltammetry - present electrodes and methods. Neuroscience, 25, 389–400.

12. Meyer, P. M., & Meyer, D. R., (1971). Neurosurgical procedures with special reference to aspiration lesions. In: Myers, R. D., (ed.), Methods in Psychobiology (Vol. 1, pp. 92–130). Academic Press, New York.

13. Miyauchi, T., Dworkin, S. I., Co, C., & Smith, J. E., (1988). Specific effects of punishment on biogenic monoamine turnover in discrete rat brain regions. Brain Res., 454, 40–50.
14. Moghaddam, B., & Bunney, B. S., (1989). Ionic composition of microdialysis perfusing solution alters the pharmacological responsiveness and basal outflow of striatal dopamine. J. Neurochem., 53, 652–654.
15. Moore, R. Y., (1978). Surgical and chemical lesion techniques. In: Iversen, L. L., Iversen, S. D., & Snyder, S. H., (eds.), Handbook of Psychopharmacology (Vol. 9, pp. 1–39). Plenum Press, New York,
16. Morley, M. J., Bradshaw, C. M., & Szabadi, E., (1987a). Effect of 6-hydroxydopamine-induced lesions of the dorsal noradrenergic bundle on steady-state operant behavior. Psychopharmacology, 93, 520–525.
17. Morley, M. J., Bradshaw, C. M., & Szabadi, E., (1987b). DSP alters the effect of d-amphetamine on variable-interval performance: Analysis in terms of Herrnstein's equation. Psychopharmacology 92, 247–253.
18. Myers, R. D., & Gurley-Orkin, L., (1985). New 'micro push-pull' catheter system for localized perfusion of diminutive structures in brain. Brain Res. Bull., 14, 1–7.
19. Myers, R. D., & Knott, P. J., (1986). Neurochemical analysis of the conscious brain: Voltammetry and push-pull perfusion. Ann. NY Acad. Sei., 473, 1–560.
20. Myers, R. D., (1972). Methods for perfusing different structures of the brain. In: Myers, R. D., (ed.), Methods in Psychobiology (Vol. 2, pp. 169–211). Academic Press, New York.
21. Myers, R. D., (1974). Handbook of Drug and Chemical Stimulation of the Brain (Vol. 1, pp. 1–22). Van Rostrand, Reinhold, NY.
22. Myers, R. D., (1977). Chronic methods: Intraventricular infusion, cerebrospinal fluid sampling, and push-pull perfusion. In: Myers, R. D., (ed.), Methods in Psychobiology (Vol. 3, pp. 281–315). Academic Press, New York.
23. Myers, R. D., Rezvani, A. H., & Gurley-Orkin, L. A., (1985). New double-lumen polyethylene cannula for push-pull perfusion of brain tissue in vivo. J. Neurosci. Methods, 12, 205–218.
24. Nielsen, J. A., & Moore, K. E., (1982). Measurement of metabolites of dopamine and 5-hydroxytryptamine in cerebroventricular perfusates of unanesthetized, freely-moving rats: Selective effects of drugs. Pharmacol. Biochem. Behav., 16, 131–137.
25. Pellegrino, L. J., & Cushman, A. J., (1971). Use of the stereotaxic

technique. In: Myers, R. D., (ed.), Methods in Psychobiology (Vol. 1, pp. 67–90). Academic Press, New York.

26. Peterson, D. W., & Sparber, S. B., (1974). Increased fixed-ratio performance and differential D- and L-amphetamine action following norepinephrine depletion by intraventricular 6-hydroxydopamine. J. Pharmacol. Exper. Ther., 191, 349–357.
27. Pettit, H. O., & Justice, J. B. Jr., (1989). Dopamine in the nucleus accumbens during cocaine self-administration as studied by in vivo microdialysis. Pharmacol. Biochem. Behav., 34, 899–904.
28. Porrino, L. J., Esposito, R. U., Seeger, T. F., Crane, A. M., Pert, A., & Sokoloff, L., (1984). Metabolic mapping of the brain during rewarding self-stimulation. Science, 224, 306–309.
29. Pycock, C. J., & Taberner, P. V., (1981). Central Neurotransmitter Turnover (Vol. 1, pp. 1–22). University Park Press, Baltimore.
30. Sakurai-Yamashita, Y., Kataoka, Y., Yamashita, K., Miyazaki, A., Ushio, M., Mine, K., Niwa, M., & Ueki, S., (1989). Conflict behavior and dynamics of monoamines of various brain nuclei in rats. Neuropharmacology, 28, 1067–1073.
31. Salamone, J. D., Keller, R. W., Zigmond, M. J., & Strieker, E. M., (1989). Behavioral activation in rats increases striatal dopamine metabolism measured by dialysis perfusion. Brain Res., 487, 215–224.
32. Sarna, G. S., Hutson, P. H., Tricklebank, M. D., & Curzon, G., (1983). Determination of brain 5-hydroxytryptamine turnover in freely moving rats using repeated sampling of cerebrospinal fluid. J. Neurochem., 40, 383–388.
33. Schallert, T., & Wilcox, R. E., (1985). Neurotransmitter-selective brain lesions. In: Boulton, A. A., & Baker, G. B., (eds.), Neuromethods: General Neurochemical Techniques (Vol. 1, pp. 343–387). Humana, Clifton, NJ.
34. Schoenfeld, R. I., & Uretsky, N. S., (1972). Operant behavior and catecholamine-containing neurons: Prolonged increase in lever-pressing after 6-hydroxydopamine. Eur. J. Pharmacol., 20, 357–362.
35. Schwarcz, R., Zaczek, R., & Coyle, J. T., (1978). Microinjection of kainic acid into the rat hippocampus. Eur. J. Pharmacol., 50, 209–220.
36. Sharp, T., Maidment, N. T., Brazell, M. P., Zetterström, T., Ungerstedt, U., Bennett, G. W., & Marsden, C. A., (1984). Changes in monoamine metabolites measured by simultaneous *in vivo* differential pulse voltammetry and intracerebral dialysis. Neuroscience, 12, 1213–1221.
37. Sharp, T., Zetterström, T., Christmanson, L., & Ungerstedt, U., (1986).

Chloroamphetamine releases both serotonin and dopamine into rat grain dialysates in vivo. Neurosci. Lett., 72, 320–324.

38. Sharp, T., Zetterström, T., Series, H. G., Carlsson, A., Grahame-Smith, D. G., & Ungerstedt, U., (1987). HPLC-EC analysis of catechols and indoles in rat brain dialysates. Life Sei., 41, 869–872.

39. Skinner, B. F., (1938). Behavior of Organisms (Vol. 1, pp. 1–22). Appleton-Century-Crofts, New York.

40. Soubrie, P., (1986). Reconciling the role of central serotonin neurons in human and animal behavior. Behav. Brain Res., 9, 319–364.

41. Sparber, S., (1975). Neurochemical changes associated with schedule-controlled behavior. Fed. Proc., 34, 1802–1812.

42. Szerb, J. E., (1967). Model experiments with Gaddum's push-pull cannulas. Can. J. Physiol.. Pharmacol. 45, 613–620.

43. Thoenen, H., & Tranzer, J. P., (1968). Chemical sympathectomy by selective destruction of adrenergic nerve endings with 6-hydroxydopamine. Nauyn-Schmiedeberg's Arch. Pharmacol., 261, 271–288.

44. Tilson, H. A., & Sparber, S. B., (1970). On the use of the push-pull cannula as a means of measuring biochemical changes during ongoing behavior. Behav. Res. Methods Instr., 2, 131–134.

45. Tye, N. C., Everitt, B. J., & Iversen, S. D., (1977). 5-Hydroxytryptamine and punishment. Nature, 268, 741–743.

46. Uhl, G. R., (1988). An approach to in situ hybridization using oligonucleotide cDNA probes. In: Van, L. F.W., Buys, R. M., Pool, C. W., & Pach, O., (eds.), Techniques in the Behavioral and Neural Sciences: Molecular Neuroanatomy (Vol. 3). Elsevier, New York.

47. Ungerstedt, U., & Hallström, Ä., (1987). In vivo microdialysis - a new approach to the analysis of neurotransmitters in the brain. Life Sei., 41, 861–864.

48. Waddington, J. L., (1986). Psychopharmacological studies in rodents: Stereotaxic intracerebral injections and behavioral assessment. In: Joseph, M. H., & Waddington, J. L., (eds.), Working Methods in Neuropsychopharmacology (Vol. 1, pp. 1–40). Manchester University Press, Manchester.

49. Yaksh, T. L., & Yamamura, H. I., (1974). Factors affecting performance of the push-pull cannula in brain. J. Appl. Physiol., 34, 428–434.

CHAPTER

8

Clinical Applications of Neurochemistry and Neurostimulation

Contents

8.1. INTRODUCTION

DBS (Deep brain stimulation) and other neurostimulation techniques have advanced dramatically in the last decade. Substantial progress has been achieved in the refinement of DBS since the technique's first introduction in the 1980s. In response to this advancement, there has been an explosion in the number of metaphors of the main mechanism of action for DBS, which includes such things as the inhibition of unusual neural activity knotted to motor control, limbic control, or cortical control, along with the activation of electrophysiological or neurochemical activities. Electrophysiological and electrochemical approaches have been developed in order to investigate these possibilities further. In order to build closed-loop neurostimulation technologies, these approaches must detect quick or immediate physiological changes. As a result, they can be employed as biomarkers. Electrophysiological approaches are employed to measure electrical potentials in order to characterize the actions of neurons. To assess changes in extracellular neurotransmitter (NT) concentrations, electrochemical studies of electro-active substances like catecholamines are performed. Even though electrophysiological techniques have received significant clinical awareness, electrochemical techniques are still in the early stages of development for clinical applications. This is because of the technical tasks of neurochemical recordings as well as the incapacity to assess long-term NT levels until relatively recently (Oh et al., 2008).

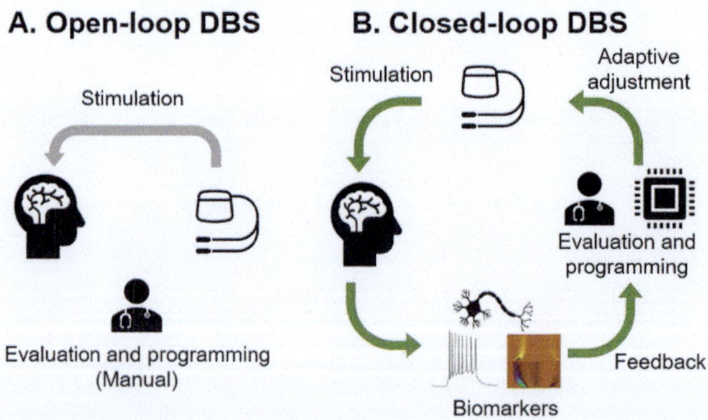

Figure 8.1. *Closed- and open-loop neurostimulation.*

Note: (A) Open-loop – the major source of input is provided by the patient, and the doctor analyzes the treatment effectiveness of the stimulator on a qualitative level. The neurostimulation settings are adjusted by the providers using an algorithm. (B) Closed-loop – the biomarker sensor provides quantitative input, whilst the patient provides qualitative feedback in the form of observations. The neurostimulation platform, under the supervision of the doctor, changes the parameters of the stimulation in real-time, resulting in feedback-sensitive and adaptive stimulation for the patient.

Source: https://thejns.org/focus/view/journals/neurosurg-focus/49/1/article-pE6.xml.

The combination of neurostimulation with subsecond neurochemical measures may usher in a new era of recording and stimulation approaches that are both more accurate and more precise. Closed-loop neurostimulation techniques, in which stimulation settings are adjusted in response to biomarker input, have the potential to simplify the process of individualizing these therapies (Figure 8.1). It is hoped that this would allow neurosurgical operations for neurostimulation-based therapies to become more up to date.

8.2. APPLICATIONS OF CLOSED-LOOP NEUROSTIMULATION IN NEUROSURGERY

Motor, behavioral, cognitive, emotional, and perceptual features that impact how people behave, feel, think, and move are all characteristics of neurological illnesses. Although the majority of people with neurological problems are effectively treated with drugs and therapy, equal to 30% of patients may not respond to traditional treatments. Deep brain simulation (DBS), which uses high-frequency electrical stimulus of cortex brain regions, is a highly effective therapy option for treatment-resistant patients with Parkinson's disease (PD) (Obeso et al., 2000; Almeida et al., 2017). Even though its clinical achievement, DBS is still linked with low effectiveness and negative therapeutic side effects. Poor manual programming of stimulation settings is partially to blame for these issues. DBS systems are currently unidirectional open-loop devices that can only offer nonadaptive stimulation based on preset settings. Continuous, time-sensitive input signals have no effect on these open-loop systems. As a result, the patient's therapeutic reaction to DBS is solely reliant on clinician evaluation, necessitating time-consuming, repetitive stimulation parameter modifications. Open-loop programming methods are severely constrained in their ability to deliver responsive, tailored stimulation because of the lack of instantaneous feedback-sensitive variation. This technological limitation is preventing optimum DBS therapy along with the clinical extension of this technology to treat complicated neurological and mental illnesses (Figure 8.2) (Little et al., 2016).

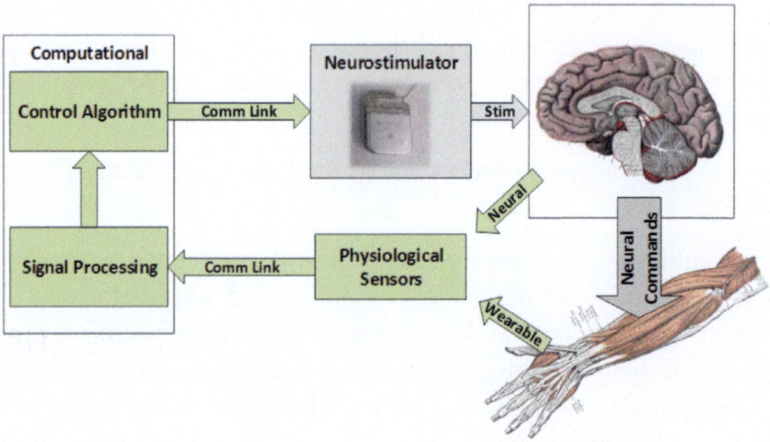

Figure 8.2. *Closed-loop through sensors and real-time communication to the implanted neurostimulator.*

Source: https://wp.ece.uw.edu/brl/neural-engineering/closed-loop-dbs/.

Biomarker objectives for feedback-sensitive neurostimulation include neurochemical compounds (Kuo et al., 2018). For decades, in vivo, electrochemical methods have been employed to detect changes in extracellular concentrations of neurochemicals like serotonin, dopamine (DA), adenosine, and glutamate in animal models with subsecond and submillimeter resolution. A variety of psychiatric and neurological disorders have been linked to abnormal neurochemical and electrophysiological activity, with relevant biomarkers for essential tremor (ET) (Marin-Lahoz, and Gironell, 2016), Parkinson's disease (PD) (Bouthour et al., 2019), TS (Tourette syndrome) (Maia and Conceição, 2018), epilepsy (Weltha et al., 2019), depression, and obsessive-compulsive disorder (OCD) (Bandelow et al., 2017; Lee et al., 2019).

8.2.1. Parkinson's Disease

The regulation of cortical brain beta and gamma oscillations is linked to the therapeutic benefits of DBS in PD (Figure 8.3) (Swann et al., 2018).

During functional neurosurgery, electrophysiological measures in basal ganglia structures, along with the oscillatory activity of LFPs in the dorsolateral motor area of the STN, have offered important understandings into the pathophysiology of PD. As a result, closed-loop DBS systems like Medtronic's Activa PC+S14 have centered on gamma and beta power as possible biomarkers for adaptive stimulation frequency and amplitude regulation (Stanslaski et al., 2012). These biomarkers direct closed-loop DBS device feedback algorithms, improving the quantifiable results of DBS. Once STN DBS was delivered only when beta power in electrophysiological activities achieved a certain limit; for example, patients with PD had a 27% improvement in motor results. Furthermore, relative to open-loop continuous stimulation, this closed-loop approach cut the overall activation time in half.

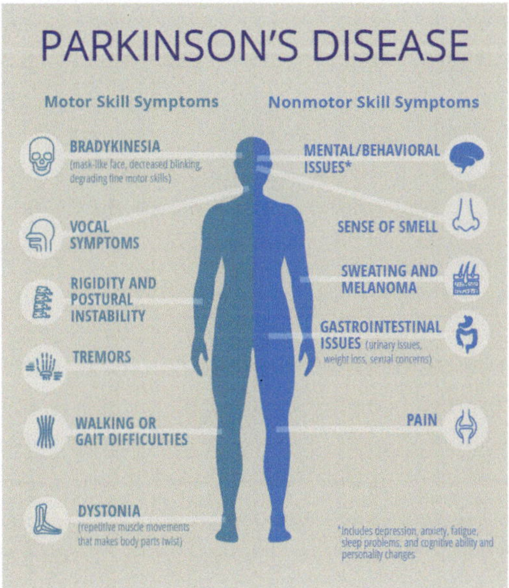

Figure 8.3. *Symptoms of Parkinson's disease.*

Source: https://www.pcori.org/research-results/pcori-stories/improving-life-women-parkinsons-disease.

Characterizing the neurochemical and electrophysiological correlates of illness symptomology is critical for the future development of closed-loop regulation of DBS for PD. DBS-induced stimulation of efferent axons and passing fibers leads to network alterations in DA neurotransmission, according to a growing body of research. The correction of aberrant oscillatory gamma and beta activity in the basal ganglia–thalamocortical electrical system may potentially underpin parkinsonian motoric relief (McIntire and Hahn, 2010). Furthermore, animal studies have indicated that brief bursts of electrical stimulation of the nucleus and STN accumbens, two prominent targets for DBS in OCD and PD patients, result in striatal phasic DA release. These findings show that striatal dopaminergic modulation may play a role in DBS's therapeutic efficacy in Parkinson's disease. DA intensity monitoring in the corpus striatum might thus serve as a possible feedback system used for closed-loop PD neurostimulation devices based on neurochemistry (Lee et al., 2006; Figee et al., 2014).

8.2.2. Essential Tremor

The participation of particular neurochemical systems in the pathogenesis of ET has not been proven, unlike DA involvement in PD. Gap junction involvement (Louis and Lenka, 2017) and serotonergic innervation in the ION (interval olivary nucleus), along with Purkinje cell loss in the cerebellar cortex, have all been proposed as possible pathophysiological origins of tremor symptoms (Ondo, 2016; Louis and Lenka, 2017). These modulatory causes, however, do not entirely account for the inhibitory and excitatory balance in oscillatory neuronal activity among the ION and the thalamus (Lin et al., 2014).

Drug therapy choices for ET and other forms of tremor have not been utilized to the point of addressing a certain neurochemical system because of a lack of information about pharmacological effects on ET. Electrophysiological indicators have therefore been predominantly employed in closed-loop DBS for tremors. In medical studies, however, our team measured FSCV (fast-scan cyclic voltammetry) in the VIM (ventral intermediate nucleus) of the thalamus in awake ET patients in response to mechanical stimulation through the implanted DBS electrode. A^3 axis accelerometer revealed a decrease in hand tremors, which corresponded to a quick, considerable rise in adenosine oxidation currents. As a result, adenosine has been studied as a possible target molecule for tremor treatment in preclinical and clinical studies. Thus, adenosine monitoring can be utilized to fine-tune stimulation levels to improve treatment success while minimizing adverse effects. However, further research into the etiology and neurochemical activities of tremors is needed (Figure 8.4) (Chang et al., 2013).

ET's pathogenesis has also been linked to anomalies in the transmission of glutamate, GABA, noradrenaline, and DA via NTs, including glutamate and noradrenaline. The greatest evidence for GABAergic impairment in the cerebellum comes from the first of these studies. GABA, on the other hand, is non-electroactive and, as a result, cannot be identified by direct electrochemical methods like amperometry or fast scanning calorimetry. The development of a dual micro biosensor for the simultaneous measurement of glutamate and GABA has recently been reported, with a detection limit for GABA of 20.12 M (Boecker et al., 2010). Research is needed to further develop methods that are more sensitive to the identification of GABA at physiological concentrations, which will be discussed later. It may also be difficult to get access to the cerebellum in order to do GABA monitoring tests. It is, therefore, necessary to conduct more research into other possible neurochemical biomarkers for the advancement of closed-loop DBS uses for ET (Hossain et al., 2018).

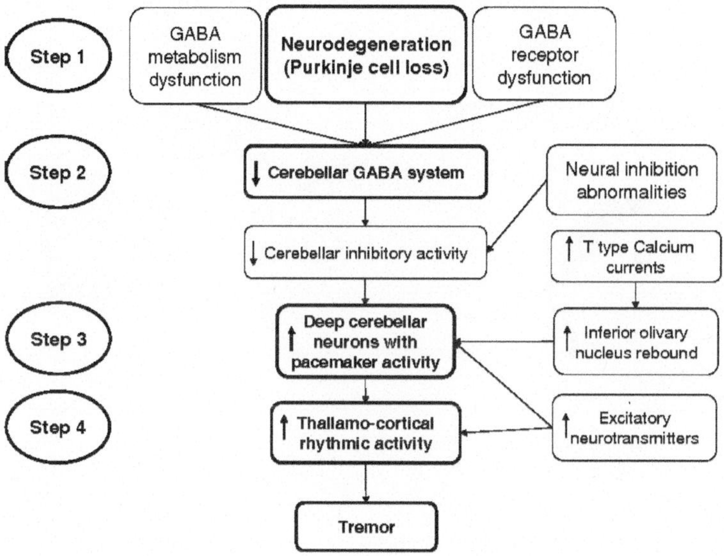

Figure 8.4. *Linking important tremor to the cerebellum.*

Source: https://www.semanticscholar.org/paper/Linking-Essential-Tremor-to-the-Cerebellum%3A-Mar%C3%ADn-Lahoz-Gironell/d65837857ac0a2ed-c9a69c6f5d0b7543d1342cdf.

8.2.3. Epilepsy

Additional potential biomarkers for epilepsy have been identified as a result of ongoing research. The Activation of the anterior nucleus of the thalamus (ANT) for Epilepsy (SANTE) experiment was one of the most important clinical tests for DBS products in epilepsy (Fisher et al., 2010). Patients exhibited a 40.4% median reduction in symptom severity following an initial blinded phase, which was encouraging. Salanova et al. (2013) showed that the median apprehension decrease had increased to 69% from baseline after 5 years. Many SANTE trial participants, on the other hand, complained of memory problems. As a result, the fornix has been proposed as an alternate target for DBS in temporal lobe epilepsy patients. Following studies used bilateral low-frequency fornix stimulation, which improved patients' hourly MMSE (mini-mental state examination) scores. Nevertheless, it was highlighted that antiepileptic medication muddled the results, and the absence of sham controls made it difficult to interpret the findings in these investigations. As a result, more research into prospective DBS targets in epilepsy patients is required (Figure 8.5) (Koubeissi et al., 2013).

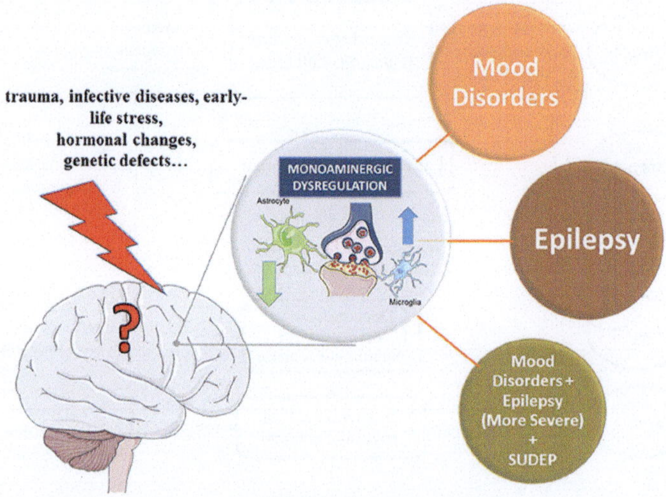

Figure 8.5. *Monoaminergic mechanisms in epilepsy.*

Source: https://www.frontiersin.org/articles/10.3389/fnins.2016.00492/full.

The establishment of a reliable disease-relevant biomarker is another stumbling block in epilepsy therapy. Specific electrophysiological activity, like interictal spikes, micro seizures, and high-frequency oscillations, is now employed as epilepsy biomarkers. Examining these electrical events in the brain allows doctors to discriminate between epileptic and non-epileptic seizures, as well as identify epileptogenic brain tissue that can be removed or targeted with DBS. Nevertheless, there has been little investigation into the use of neurochemicals to monitor disease states. Extracellular glutamate concentrations at the seizure location are elevated during or after seizures, according to recent microdialysis investigations, indicating a probable function in seizure-related brain structures (Çavuş et al., 2016). This discovery suggests that after implanting both glutamate and stimulation-recording electrodes into seizure locations, feedback control may be conceivable. However, further research is needed to determine the specific pathways through which neurochemical substances lead to seizure activity.

8.2.4. Tourette Syndrome

Tics are a type of neurodevelopmental condition that causes motor and verbal tics. However, TS is generally a childhood disorder that increases over time; a small percentage of people may not react to treatment and maintain to have symptoms as adults. For many seriously afflicted individuals, DBS has emerged as a viable therapy option. The thalamic centromedian-parafascicular complex and the globus pallidus internal (GPi) have been identified as two key basal ganglia targets for DBS treatment (CM-Pf) (Figure 8.6) (Xu et al., 2020).

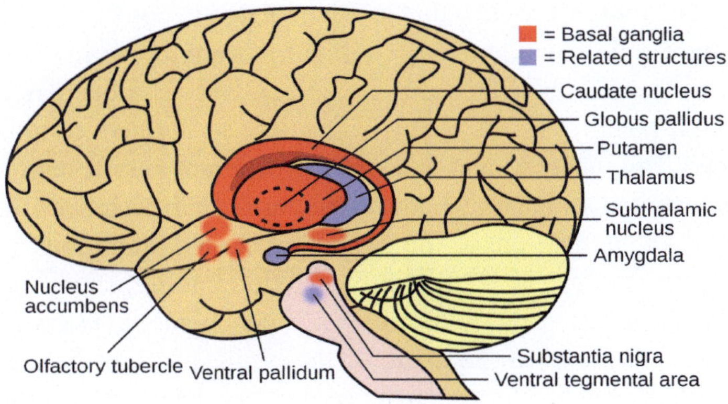

Figure 8.6. *The basal ganglia and thalamus are implicated in Tourette syndrome.*

Source: https://en.wikipedia.org/wiki/Tourette_syndrome.

There have been approximately 180 individuals treated with DBS for TS thus far, with a significant decrease in tics. According to the Yale Global Tic Severity Scale, CM-Pf DBS reduced motor tics by 46% and phonic tics by 52% in 10 individuals with resistant TS (YGTSS). Targeting the GPi has also been shown to be effective in studies. According to the YGTSS, a new double-blind crossover experiment of 14 individuals, found a 15% decrease in tic intensity. Furthermore, a recent multisite retrospective investigation demonstrated that both CM-Pf and GPi stimulation resulted in a significant increase in YGTSS scores, with no variations in treatment results across targets (Testini et al., 2016).

In spite of these accomplishments, the therapeutic method through which DBS works remains a mystery. The primary emphasis of research has been on the effects of DBS on neurotransmission and neurochemical release. It's been suggested that DBS impacts circuits terminating in the striatum, depending on the stimulation aim. Increased activity in alpha bands and low-frequency was discovered in LFP recordings in the GPi and thalamus, supporting this assertion. It's also been looked at utilizing LFP recordings from the CM-Pf through stimulation in a feed-forward approach to adjust DBS settings. As a result, LFP recordings might be useful biomarkers for closed-loop stimulation stages.

8.2.5. Chronic Pain

A total of 20.4% of individuals in the United States suffer from chronic pain, which is defined as pain that lasts longer than the average period of damage and healing. Chronic pain has been related to opioid use, along with the subsequent opioid pandemic, according to recent research. Furthermore, pain adds to feelings of worry and despair, as well as a decreased overall level of well-being. Chronic pain is not a well-defined condition with well-defined pathogenesis, and it has many different manifestations. It has been demonstrated in a large body of research that peripheral or central neurological slights can cause long-lasting diseased alterations in several brain areas that are engaged in pain conduction and transmission. System for processing somatosensory, emotional, and cognitive information can be distinguished among these areas of the brain. It might be challenging to achieve appropriate chronic pain control by behavioral treatment or medication because of the multidimensionality of the pain experience. As a result, deep brain stimulation (DBS) has been identified as a possible treatment alternative in refractory patients (Figure 8.7) (Martinez-Ramirez et al., 2018).

Since the first trials of deep brain stimulation (DBS) for pain were conducted, several investigations for a wide range of pain symptoms have been conducted. The periaqueductal gray (PAG) and periventricular gray matter (PVG) regions, as well as the vT (ventral thalamus), which includes the ventral posteromedial nucleus (VPM) and ventral posterolateral nucleus (VPL), the ventral striatum (VS), and the anterior cingulate cortex (ACC) and frontal limb of the internal capsule, were the most frequently identified DBS targets in these studies (ALIC). A total of over 200 patients with a range of chronic pain symptoms were included in the world's first multicenter randomized controlled study of DBS for pain. DBS procedures were carried out in either the vT or the PAG. This experiment, on the other hand, failed to reveal a statistically meaningful improvement. Successive studies have been carried out with varying degrees of success. Boccard et al. conducted a successful trial in which 74 patients were treated with VPL/VPM (vT) and/or PAG/PVG stimulation for one year and saw a mean pain reduction of 46%. 48 Another trial conducted by Boccard et al. that targeted the ACC reported a 43.4% reduction in the number of patients after one year in 22 patients (Testini et al., 2016).

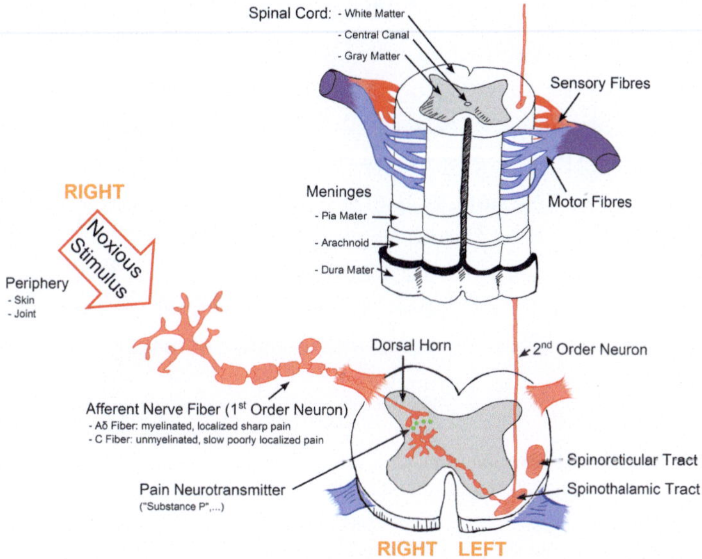

Figure 8.7. *Representation of the pain pathway (neurotransmission), showing the different afferent fibers and neurotransmitter release within the dorsal horn.*

Source: https://teachmephysiology.com/nervous-system/sensory-system/pain-pathways/.

DBS feedback control is currently being investigated, although no acceptable biomarkers for this purpose have yet been found. GABA, DA, glutamate, and serotonin are all potential molecular biomarkers for pain, medicines that influence these NT systems are now being employed in the treatment of chronic pain. As a result of its compatibility with neurochemical measuring methods, DA may have the most potential for use in pain management monitoring; in addition, multiple imaging investigations have revealed abnormal striatal DA activity in pain patients. In addition, the use of LFP recording has been suggested. New research by Shirvalkar et al. suggested that recording of oscillations in the S1 primary sensory cortex as well as areas of the orbitofrontal cortex (OFC) and the ACC could be used to expect chronic pain as well as to adjust stimulation parameters for patients with fibromyalgia (fibromyalgia Nervosa). Given the wide variety of potential biomarkers, along with the varied and unexpected responses to

DBS, it is probable that a closed-loop DBS system along with fluid stricture adjustment may be required to generate a dependable therapeutic response in chronic pain patients (Figure 8.8) (Martínez-Fernández et al., 2011).

8.2.6. Obsessive-Compulsive Disorder (OCD)

Since the initial deep brain stimulation (DBS) studies for pain were carried out, a large number of investigations have been carried out for a wide range of pain indications. DBS goals in these studies included the PVG and periaqueductal gray (PAG) regions, as well as the ventral thalamus (vT), which contains the VPM and ventral posterolateral nucleus (VPL), the ventral striatum (VS), the ACC, and the anterior limb of the internal capsule (ALIC). This was the world's first multicenter randomized controlled research using deep brain stimulation (DBS) for pain, which included more than 200 patients with a number of chronic pain indications. In all, DBS surgeries were performed in either the PAG or the vT setting. This trial, on the other hand, was unable to demonstrate a statistically significant improvement in performance.

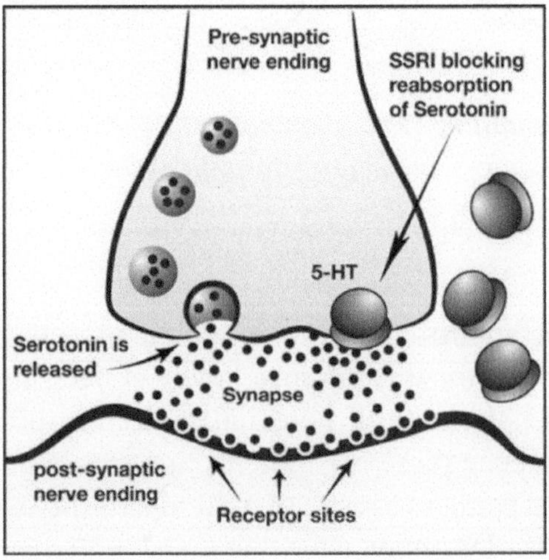

Figure 8.8. *Neurotransmitters and receptive sites in obsessive-compulsive disorder.*

Source: https://www.ocduk.org/ocd/what-causes-ocd/.

Further research has been carried out, with varying degrees of success, in the years since. Although no appropriate biomarkers for DBS response control have yet been discovered, researchers are now investigating the possibility of using them. Molecular biomarkers for pain such as GABA,

DA, glutamate, and serotonin have all been identified as possible molecular biomarkers for pain. This is because medications that modulate these NT systems are now being used in the treatment of chronic pain. DA may present the most potential for application in pain management monitoring because of its compatibility with neurochemical measurement techniques; in addition, several imaging investigations have demonstrated aberrant striatal DA activity in pain sufferers. The use of LFP recording has also been suggested as an alternative.

8.2.7. Depression

The production of closed-loop neurostimulation systems faces a difficult problem in the case of depression. This is owing in part to the ongoing dispute about the best stimulation targets, including the scarcity of accurate biomarkers together with high disease specificity. Initial trials found therapeutic benefit when white matter tracts adjacent to the subgenual cortex region and the VC/VS were stimulated. Nevertheless, following multicenter randomized controlled studies in both BA2565 and VC/VS66 failed to discriminate sham from stimulation, lowering initial excitement for DBS for depression (Figure 8.9).

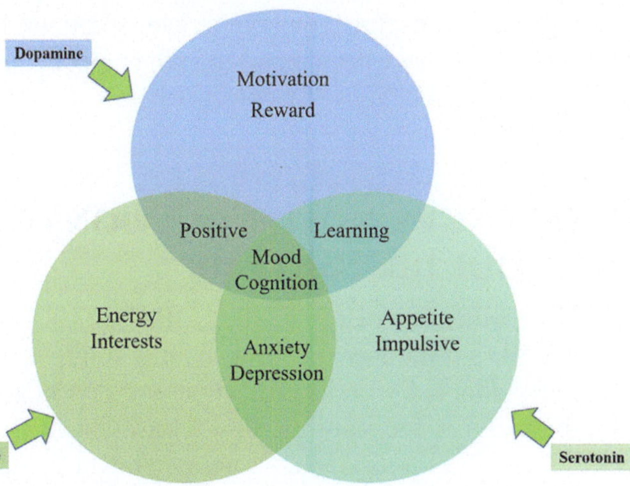

Figure 8.9. *Effects of neurochemical transmission on mood variation.*

Source: https://www.frontiersin.org/articles/10.3389/fpsyg.2018.02201/full.

Though initially disheartening, further research has revealed that modifications in study design, probe location, and stimulation settings may have had a significant influence on the findings of these experiments. For example, following probe insertion, studies frequently evaluate sham and stimulation findings before providing open-label therapy. However, independent of stimulation, the provocative response to probe implantation has been shown to have therapeutic effects, posing a serious confounder to this research design. An alternate approach may give all patients with open-label therapy before blinded discontinuance at a predetermined time point, rather than making comparisons the therapeutic advantages of sham versus stimulation following electrode placement (Little et al., 2013).

DBS has a lot of potential as a therapy for depression that hasn't responded to other treatments. To realize this potential, however, further research into the therapeutic mechanism of DBS is needed, as well as more fine-tuning of the stimulation settings. To optimize closed-loop systems, biomarkers must be carefully chosen to provide feedback for adaptive stimulation settings. Biomarker selection, however, is difficult due to the extremely multimodal and idiosyncratic character of depression. Individual pre-testing of unique depression pathophysiology to identify appropriate stimulation settings will very certainly differ between manifestations and subtypes of depression, requiring individual pre-testing of particular depression pathophysiology. Given their involvement in depression neurochemistry and therapy, DA, and serotonin may be useful biomarkers. Modifications in serotonin and DA neurotransmission have been reported in the hippocampus and striatum, indicating plausible locations for electrochemical nursing (Maia and Conceição, 2018).

8.3. CHALLENGES AND INNOVATIONS OF CLOSED-LOOP DBS SYSTEMS

However, early indicators of closed-loop DBS' efficacy are encouraging. The technology has a number of obstacles that must be overcome before it can be widely adopted. The identification and dependable measurement of suitable biomarkers for real-time illness assessment is a key hurdle to the establishment of closed-looped systems. However, no one biomarker is adequate for tracking the evolution of neurological illness symptoms or DBS applications; further study is needed to extract biomarkers and link them to disease states and unique pathophysiologies. As a result, a closed-loop system that can concurrently monitor numerous biomarkers with known disease significance may be able to accomplish more granular disease monitoring.

REFERENCES

1. Almeida, L., Deeb, W., Spears, C., Opri, E., Molina, R., Martinez-Ramirez, D., & Okun, M. S., (2017). Current practice and the future of deep brain stimulation therapy in Parkinson's disease. In: *Seminars in Neurology* (Vol. 37, No. 2, pp. 205–214). Thieme Medical Publishers.
2. Bandelow, B., Baldwin, D., Abelli, M., Bolea-Alamanac, B., Bourin, M., Chamberlain, S. R., & Riederer, P., (2017). Biological markers for anxiety disorders, OCD and PTSD: A consensus statement. Part II: Neurochemistry, neurophysiology and neurocognition. *The World Journal of Biological Psychiatry, 18*(3), 162–214.
3. Boecker, H., Weindl, A., Brooks, D. J., Ceballos-Baumann, A. O., Liedtke, C., Miederer, M., & Miederer, I., (2010). GABAergic dysfunction in essential tremor: An 11C-flumazenil PET study. *Journal of Nuclear Medicine, 51*(7), 1030–1035.
4. Bouthour, W., Mégevand, P., Donoghue, J., Lüscher, C., Birbaumer, N., & Krack, P., (2019). Biomarkers for closed-loop deep brain stimulation in Parkinson's disease and beyond. *Nature Reviews Neurology, 15*(6), 343–352.
5. Çavuş, I., Romanyshyn, J. C., Kennard, J. T., Farooque, P., Williamson, A., Eid, T., & Spencer, D. D., (2016). Elevated basal glutamate and unchanged glutamine and GABA in refractory epilepsy: Microdialysis study of 79 patients at the yale epilepsy surgery program. *Annals of Neurology, 80*(1), 35–45.
6. Chang, S. Y., Kim, I., Marsh, M. P., Jang, D. P., Hwang, S. C., Van, G. J. J., & Lee, K. H., (2012). Wireless fast-scan cyclic voltammetry to monitor adenosine in patients with essential tremor during deep brain stimulation. In: *Mayo Clinic Proceedings* (Vol. 87, No. 8, pp. 760–765). Elsevier.
7. Chang, S. Y., Kimble, C. J., Kim, I., Paek, S. B., Kressin, K. R., Boesche, J. B., & Lee, K. H., (2013). Development of the mayo investigational neuromodulation control system: Toward a closed-loop electrochemical feedback system for deep brain stimulation. *Journal of Neurosurgery, 119*(6), 1556–1565.
8. Figee, M., De Koning, P., Klaassen, S., Vulink, N., Mantione, M., Van, D. M. P., & Denys, D., (2014). Deep brain stimulation induces striatal dopamine release in obsessive-compulsive disorder. *Biological Psychiatry, 75*(8), 647–652.
9. Fisher, R., Salanova, V., Witt, T., Worth, R., Henry, T., Gross, R., & SANTE Study Group, (2010). Electrical stimulation of the anterior nucleus of thalamus for treatment of refractory epilepsy. *Epilepsia, 51*(5), 899–908.

10. Hossain, I., Tan, C., Doughty, P. T., Dutta, G., Murray, T. A., Siddiqui, S., & Arumugam, P. U., (2018). A novel microbiosensor microarray for continuous ex vivo monitoring of gamma-aminobutyric acid in real-time. *Frontiers in Neuroscience, 12*, 500.
11. Koubeissi, M. Z., Kahriman, E., Syed, T. U., Miller, J., & Durand, D. M., (2013). Low-frequency electrical stimulation of a fiber tract in temporal lobe epilepsy. *Annals of Neurology, 74*(2), 223–231.
12. Kuo, C. H., White-Dzuro, G. A., & Ko, A. L., (2018). Approaches to closed-loop deep brain stimulation for movement disorders. *Neurosurgical Focus, 45*(2), E2.
13. Lee, E. H., & Han, P. L., (2019). Reciprocal interactions across and within multiple levels of monoamine and cortico-limbic systems in stress-induced depression: A systematic review. *Neuroscience & Biobehavioral Reviews, 101*, 13–31.
14. Lee, K. H., Blaha, C. D., Harris, B. T., Cooper, S., Hitti, F. L., Leiter, J. C., & Kim, U., (2006). Dopamine efflux in the rat striatum evoked by electrical stimulation of the subthalamic nucleus: Potential mechanism of action in Parkinson's disease. *European Journal of Neuroscience, 23*(4), 1005–1014.
15. Lin, C. Y., Louis, E. D., Faust, P. L., Koeppen, A. H., Vonsattel, J. P. G., & Kuo, S. H., (2014). Abnormal climbing fibre-Purkinje cell synaptic connections in the essential tremor cerebellum. *Brain, 137*(12), 3149–3159.
16. Little, S., Beudel, M., Zrinzo, L., Foltynie, T., Limousin, P., Hariz, M., & Brown, P., (2016). Bilateral adaptive deep brain stimulation is effective in Parkinson's disease. *Journal of Neurology, Neurosurgery & Psychiatry, 87*(7), 717–721.
17. Little, S., Pogosyan, A., Neal, S., Zavala, B., Zrinzo, L., Hariz, M., & Brown, P., (2013). Adaptive deep brain stimulation in advanced Parkinson disease. *Annals of Neurology, 74*(3), 449–457.
18. Louis, E. D., & Lenka, A., (2017). The olivary hypothesis of essential tremor: Time to lay this model to rest?. *Tremor and Other Hyperkinetic Movements, 7*.
19. Maia, T. V., & Conceição, V. A., (2018). Dopaminergic disturbances in Tourette syndrome: An integrative account. *Biological Psychiatry, 84*(5), 332–344.
20. Marin-Lahoz, J., & Gironell, A., (2016). Linking essential tremor to the cerebellum: Neurochemical evidence. *The Cerebellum, 15*(3), 243–252.
21. Martínez-Fernández, R., Zrinzo, L., Aviles-Olmos, I., Hariz, M., Martinez-

Torres, I., Joyce, E., & Foltynie, T., (2011). Deep brain stimulation for Gilles de la Tourette syndrome: A case series targeting subregions of the globus pallidus internus. *Movement Disorders, 26*(10), 1922–1930.

22. Martinez-Ramirez, D., Jimenez-Shahed, J., Leckman, J. F., Porta, M., Servello, D., Meng, F. G., & Okun, M. S., (2018). Efficacy and safety of deep brain stimulation in Tourette syndrome: The international Tourette syndrome deep brain stimulation public database and registry. *JAMA Neurology, 75*(3), 353–359.
23. McIntyre, C. C., & Hahn, P. J., (2010). Network perspectives on the mechanisms of deep brain stimulation. *Neurobiology of Disease, 38*(3), 329–337.
24. Obeso, J. A., Olanow, C. W., & Nutt, J. G., (2000). Levodopa motor complications in Parkinson's disease. *Trends in Neurosciences, 23*, S2–S7.
25. Oh, Y., Heien, M. L., Park, C., Kang, Y. M., Kim, J., Boschen, S. L., & Jang, D. P., (2018). Tracking tonic dopamine levels *in vivo* using multiple cyclic square wave voltammetry. *Biosensors and Bioelectronics, 121*, 174–182.
26. Ondo, W., (2016). Essential tremor: What we can learn from current pharmacotherapy. *Tremor and Other Hyperkinetic Movements, 6*.
27. Salanova, V., Witt, T., Worth, R., Henry, T. R., Gross, R. E., Nazzaro, J. M., & SANTE Study Group, (2015). Long-term efficacy and safety of thalamic stimulation for drug-resistant partial epilepsy. *Neurology, 84*(10), 1017–1025.
28. Stanslaski, S., Afshar, P., Cong, P., Giftakis, J., Stypulkowski, P., Carlson, D., & Denison, T., (2012). Design and validation of a fully implantable, chronic, closed-loop neuromodulation device with concurrent sensing and stimulation. *IEEE Transactions on Neural Systems and Rehabilitation Engineering, 20*(4), 410–421.
29. Swann, N. C., De Hemptinne, C., Thompson, M. C., Miocinovic, S., Miller, A. M., Ostrem, J. L., & Starr, P. A., (2018). Adaptive deep brain stimulation for Parkinson's disease using motor cortex sensing. *Journal of Neural Engineering, 15*(4), 046006.
30. Testini, P., Min, H. K., Bashir, A., & Lee, K. H., (2016). Deep brain stimulation for Tourette's syndrome: The case for targeting the thalamic centromedian–parafascicular complex. *Frontiers in Neurology, 7*, 193.
31. Testini, P., Zhao, C. Z., Stead, M., Duffy, P. S., Klassen, B. T., & Lee, K. H., (2016). Centromedian-parafascicular complex deep brain stimulation for Tourette syndrome: A retrospective study. In: *Mayo Clinic Proceedings* (Vol. 91, No. 2, pp. 218–225). Elsevier.

32. Weltha, L., Reemmer, J., & Boison, D., (2019). The role of adenosine in epilepsy. *Brain Research Bulletin, 151*, 46–54.
33. Xu, W., Zhang, C., Deeb, W., Patel, B., Wu, Y., Voon, V., & Sun, B., (2020). Deep brain stimulation for Tourette's syndrome. *Translational Neurodegeneration, 9*(1), 1–19.

INDEX

E

F

G

H

O

P

Q

R

S